GROWING COMMUNITY FORESTS

GROWING COMMUNITY FORESTS

Practice, Research, and Advocacy in Canada

Edited by Ryan Bullock, Gayle Broad, Lynn Palmer, and M.A. (Peggy) Smith

UNIVERSITY OF MANITOBA PRESS

Growing Community Forests: Practice, Research, and Advocacy in Canada

21 20 19 18 17 1 2 3 4 5

University of Manitoba Press
Winnipeg, Manitoba, Canada
Treaty 1 Territory
uofmpress.ca

Cataloguing data available from Library and Archives Canada
ISBN 978-0-88755-793-4 (PAPER)
ISBN 978-0-88755-533-6 (PDF)
ISBN 978-0-88755-531-2 (EPUB)

Cover design by Sébastien Aubin
Interior design by Karen Armstrong

Printed in Canada

This book has been published with the help of a grant from the Federation for the Humanities and Social Sciences, through the Awards to Scholarly Publications Program, using funds provided by the Social Sciences and Humanities Research Council of Canada.

The University of Manitoba Press acknowledges the financial support for its publication program provided by the Government of Canada through the Canada Book Fund, the Canada Council for the Arts, the Manitoba Department of Sport, Culture, and Heritage, the Manitoba Arts Council, and the Manitoba Book Publishing Tax Credit.

Funded by the Government of Canada | Canada

Contents

PART 2: Fostering Community Capacity, Enterprise, and Diversification

List of Illustrations

Figures

Tables

INTRODUCTION

Growing Community Forests

Ryan Bullock, Gayle Broad, Lynn Palmer, and M.A. (Peggy) Smith

Canada is slowly emerging from an unparalleled crisis involving forests and communities across the country. The Canadian forest sector crash of 2007–8 should have served as a "canary in the coal mine" for the 2008–9 world economic crisis. What first appeared as a typical sector downturn associated with a drop in U.S. demand for Canadian wood products quickly became known as the "perfect storm" north of the U.S.–Canada border, where a combination of factors precipitated a crisis that surpassed the experiences and memories of even very senior forest managers (Bullock et al. 2015). Framed in economic terms, the crisis involved major decreases in conventional wood product markets, which had significant and direct impacts on employment across mill and woodland operations. Between 2005 and 2015, direct jobs decreased from 316,850 to 201,645, respectively, for an overall 36 percent decrease (NRCAN 2016). During this period, forestry's Gross Domestic Product (GDP) contribution decreased by almost half, from $30 billion to approximately $17 billion, before recovering to $22 billion in 2015 (NRCAN 2016). The crisis exhibited many long-recognized problems associated with natural resource industry closures in single-industry towns (Lucas 1971). Notable changes included widespread unemployment but also forced regional migration and the loss of skilled labour and youth, and the loss of municipal services and programs due to dwindling local tax bases. However, the wave of permanent mill closures and layoffs led to industry and government reorganization, provincial policy reforms, and new municipal, provincial, and federal strategies, as well as Indigenous and labour initiatives—all efforts to help navigate the transition through what turned out to be an unprecedented period of turbulence and conflict for the Canadian forest sector and forest peoples.

While communities, policy makers, and industry leaders, among others, acknowledge common economic challenges, such as an overdependence on foreign markets, rising energy costs, and lack of diversification, no unified set of solutions has been implemented. What is more, addressing the complex set of social and ecological factors related to changes in forest governance and development presents a unique challenge for decision makers and citizens of Indigenous and non-Indigenous communities that depend upon forest lands and resources. Our dependence and influence on forest ecosystems becomes all-important when the connections among livelihoods, as well as identities and cultures, are fully recognized. Ongoing and at times contentious public debate has revealed an appetite for fundamentally rethinking relationships that link communities, governments, industries, and forests in a movement toward a more sustainable future.

Among the paths possible, the creation of community forests, and similar collaborative governance and development arrangements, promises to strengthen forest communities and ecosystems. It has long been argued that increased local control over common forest lands can help activate resource development opportunities, benefits, and stewardship responsibilities (Dunster 1989; Hammond 1990; Charnley and Poe 2007). But implementing community forestry has proven to be a complex task (see Furness et al. 2015; Ambus and Hoberg 2011; Bullock et al. 2009; Duinker et al. 1991). As M'Gonigle and Dempsey (2003: 116) point out, "there are dangers associated with naive confidence in community forests." Institutions and plans must be mindfully constructed to ensure that equitable control and benefit sharing can be achieved. Still, there is a rich and growing body of experience to draw upon in settings where community forests have taken hold. There is also more research being done with community forest initiatives in Canada than in the past (Bullock and Lawler 2015). National and regional efforts to coordinate and share resources and local stories to support community forests remain in their infancy, but this is changing. This edited volume is intended to help address the need to bridge practice, research, and advocacy for communities and their forests.

This introduction provides an overview of forest–community linkages, and it sets out the main concepts, issues, and context needed to understand the evolution of community forests in Canada. The next section of this introduction presents important considerations and points of debate regarding terminology and concepts. We then briefly review the state of domestic forest policy and community forest practice, research, and advocacy, with selected reference to global trends and examples. Finally, an overview of the structure of the book is provided.

Clarifying Terminology for "Community Forestry" in Canada

While community forestry is an age-old practice, it received minimal scholarly attention in Canada prior to the mid-1980s. Notably, Marshall (1986) appears to have first described "community forest licences" in British Columbia in his paper published in *Forest Planning Canada*. Shortly thereafter, the first papers describing community forestry practices (e.g., Roy 1989) and concepts (e.g., Dunster 1989) were published. These were mainly attempts to explore, articulate, and illustrate what community forestry was and how it could work. Soon after these early works, Duinker and colleagues (1994: 711) observed that there was growing interest and debate among various groups: "Community forests, community forestry—forest managers and policy makers in Canada today cannot escape these concepts." Scholarship on the topic has since steadily increased, engaging a growing list of themes and keywords, and scholarly debate on appropriate terminology continues to evolve (Bullock and Lawler 2015; Teitelbaum 2016). It must not be ignored that decades of grassroots work have shaped community forestry concepts and practice in Canada as well (see Booth 1998; Pinkerton 1993).

A main debate is whether "community forest" and "community forestry" should be used interchangeably, though they often are (see, e.g., Duinker et al. 1991; Bullock and Hanna 2012; Benner et al. 2014; Furness et al. 2015). The meanings of these terms are interpreted quite broadly and sometimes even used in conjunction with related terminology (discussed below). In considering a range of definitions and uses, a main distinction can be made between:

(a) *community forest as the site* where management goals and initiatives are directed and carried out to produce benefits, and

(b) *community forestry as the process and practice* in which local values, objectives, and decisions are developed and implemented to mobilize joint actions affecting forest ecosystems and communities.

For example, Benner et al. (2014) use the two different terms quite effectively in this manner. With respect to the frequency of use, recent bibliometrics research indicates that "community forestry" is used nearly twice as often as "community forest" as an author-selected keyword in research journals (Bullock and Lawler 2015). It is important to note that many decades earlier "community forest"—in practice and concept—existed under many other aliases, such as "municipal woodlot" (MacDonald 1935), "county forest" (Cain 1939), "town forest" (Wilson 1943), and "forest village" (Auden 1944). Each referred to a forest(s) managed by a local entity, most of which had emerged from government conservation programs and what can be described as regional economic stimulus initiatives (i.e., job creation and settlement programs). So despite the fact that, in general, "definitions and terms for community forestry abound in the literature" (Charnley and Poe 2007: 303), it would seem that "community forestry" is emerging as the most widely used and preferred term, in the Canadian context at least.

Preferences for specific language aside, scholars highlight three core aspects that encompass a broader set of issues related to community forestry: (1) local decision making, (2) local values, and (3) diverse local benefits (Teitelbaum 2014; Bullock and Hanna 2012; Charnley and Poe 2007; McCarthy 2005) (Table 0.1). These aspects are integrated by Charnley and Poe (2007: 303), who explain that "in community forestry, (*a*) some degree of responsibility and authority for forest management is formally vested by the government in local communities; (*b*) a central objective of forest management is to provide local communities with social and economic benefits from forests; and (*c*) ecologically sustainable forest use is a central management goal, with forest communities taking some responsibility for maintaining and restoring forest health."

Table 0.1. Core aspects of community forestry.

Aspect	Description	Examples
Decision making	Authority and ability for decision making in forest planning, management, and development resides at the local level; forest community representatives empowered to decide on various normative, strategic, and operational matters.	• Design governance structures/ processes • Identify stakeholders • Set harvest levels • Identify where/when harvesting occurs • Determine who will undertake harvesting • Direct wood flow to end use

Aspect	Description	Examples
Local values	Social interactions, forest conditions, and forestry products deemed important by and for the community, whether for sociocultural, economic, or ecological reasons are upheld	• Timber • Non-timber forest products • Sacred sites • Education • Habitat • Scenic views
Local benefits	Human interactions with forests foster physical, emotional, and spiritual gains, which are felt at individual and community level.	• Local/regional employment • Learning and training opportunities • Community service and program provision • Maintain cultural identities and ways of life • Self-reliance, capacity, and autonomy • Maintenance of healthy ecosystem

As research on community forestry and the terminology used to describe it has evolved, other terms have entered the lexicon, some of which are commonly associated with, and at times have been used interchangeably with, community forestry, despite some possible differences in intended meaning and practice (see Krogman and Beckley 2002). For starters, several recent authors have used "community-based forestry" (Alemagi 2010), "community-based forest management" (e.g., Robson 2014; Kube 2012; Charnley and Poe 2007), and the more general "community-based natural resource management" (e.g., Bixler 2014; Ambus and Hoberg 2011; Bradshaw 2003) to describe governance arrangements aimed at increasing both public involvement and the variety and number of forest benefits and interests involved. Aboriginal forestry (Parsons and Prest 2003), First Nations forestry (Wyatt 2008), and Indigenous community forests (Belsky 2008) intuitively seem to fit with many of the economic, social, and ecological considerations of community forestry as we know it today. These terms have also included wide-reaching sets of concepts that set out various governance arrangements for Indigenous inclusion based on traditional ways and knowledge (Wyatt 2008).

Utilization of different terms to describe the same phenomenon, or the converse, appears to be driven by personal preference but also the variability of community involvement in forest management processes and enterprises and the consequent vagueness of related concepts. There are now several approaches to natural resource management that sometimes get associated with community forestry, including co-management, joint ventures, and public advisory boards. Often these labels are used loosely and, admittedly, it can sometimes be difficult to distinguish among examples. In Table 0.2 below we summarize how some of these terms are described and exemplified in relevant Canadian literature.

Table 0.2. Descriptions of terminology frequently used in conjunction with community forestry.

Term	Description	Examples
Co-management	An organizational model that involves sharing decision-making responsibilities and usually combines a government management style with a local, traditional, consensus-based style. Often occurs in Canada between Indigenous communities and provincial or territorial governments (Beckley 1998).	• John Prince Research Forest (Grainger et al. 2006) • The Tin Wis Coalition (Pinkerton 1993) • Clayoquot Sound (Mabee and Hoberg 2006)
Joint Venture	A business relationship between a community and a forestry company intended to build working relationships and explore new economic opportunities. It can often be difficult to reflect differing cultural values through joint venture agreements (Brubacher 1998).	• Cree of Waswanipi, Quebec, and Domtar Inc. • West Chilcotin Forest Products, Anahim Lake, BC (Brubacher 1998)

Public Advisory Board	Also known as stakeholder advisory committees, these are tools used to increase public participation and influence over policy. An advisory board represents and communicates the ideas and values of a community throughout the decision-making process of a proposal (McGurk et al. 2006).	• Louisiana-Pacific's Stakeholder Advisory Committee (Parkins et al. 2006) • Various Local Citizen Committees in northern Ontario (Robson 2014)

Community forestry is sometimes equated with government- and industry-led public participation processes (see Robson 2014). While many organizations, including some community forests, use advisory committees as one way to engage citizens in decision making and other activities (see Teitelbaum and Bullock 2012), the actual level of involvement achieved through such activities does not fully meet the community forestry ideal. Likewise, community forest organizations are, on occasion, associated with other forest-oriented organizations and multi-stakeholder arrangements intended to elevate and strengthen participation of a broad range of groups and individuals in, for example, forest research and education, regional discussion forums, and entrepreneurial endeavours. This would include likening community forests to model forests (e.g., Barkley et al. 1997) and other research or demonstration forests (e.g., Grainger et al. 2006). While such "experimental" forests can operate under organizations with similar value systems, their activities are usually specifically focused on research, stakeholder networking, knowledge exchange, and other forestry extension activities. As illustrated in the chapters to come, community forest initiatives endeavour to decentralize and devolve decision making (see Ribot 2002, 2004; Berkes 2010). Under such arrangements, local representatives who are downwardly accountable to their constituents collectively make decisions about the use and control of common forests that are important to communities. To do this, community forests, and efforts to support them in Canada, engage in and integrate practice and policy with research and advocacy. Below we briefly outline key advances in these domains within the Canadian context.

Policy, Practice, and Forest Communities

Canadian forest policy changes of most relevance to forest communities are linked to increasing public concerns about the impacts of management decisions and the locus of decision-making control, forest ownership, and the primary beneficiaries of forest resource development (Beckley 1998; Haley and Nelson 2007; Bullock and Hanna 2012). Historically, under the conventional command-and-control model of forest policy making and management, there was no place for communities beyond that of providing labour and services to forest companies that extracted public natural resources for private profit and, in turn, provincial revenue (see Bullock and Reed 2016). From a regional development point of view, the goal was to create access to public forests to attract private investment in order to encourage development of resources and infrastructure, employment, and population growth in settler communities (Hayter and Barnes 1997). Little to no consideration was given to Indigenous development.

From a policy-making perspective, the provinces had jurisdiction over natural resources and so made policies to support emerging resource sectors. Large bureaucracies and closed policy networks consisting of government and industry actors formed to administer large-scale forest development (Ross 1995; Hessing et al. 2005). With the exception of the "appurtenancy" requirements (i.e., local processing of logs) meant to ensure economic spin-offs from forestry within Canada (Hickey and Nelson 2005), community development and planning were not initially primary considerations for natural resource ministries and companies. This policy outlook persisted for much of Canada's industrial forestry history. However, there is now growing acknowledgement of the direct links between communities and their forests, reinforced by the fact that approximately 94 percent of Canada's forests are on public land (NRCAN 2014).

Since the early boom years of the conservationist era (beginning c. 1900) a permanent forest industry and private forest enterprise have been considered essential to economic growth (Apsey et al. 2000; Beckley 2003). Despite the fact that the majority of commercially productive forests are located in places where there is a high degree of community dependence on publicly owned resources, forests have remained controlled by provincial natural resource departments and the forest industry through what is known as the "tenure system." The current tenure system is rooted in the colonial ideology of the mid-nineteenth century, when foundational policies and legislation were designed to (1) retain Crown ownership and control of forest lands and resources, (2) promote industrial extraction and sale of timber from public lands, and (3) create provincial revenues from Crown timber sales (Drushka 2003; Nelles 2005). In addition to creating a long-standing provincial government dependence on Crown timber revenues, "these policies formed the basis of a unique Canadian partnership between timber-based forest industries and colonial (later provincial) governments that controlled the majority of the forest land" (Drushka 2003: 30).

Canada's current tenure policies date back to the 1940s post–Second World War era and are based on the notion of sustained yield harvesting, which was meant to ensure steady fibre flows to designated wood-processing facilities through large-scale timber extraction (Burton et al. 2003; Haley and Nelson 2007). Founded on neoclassical resource economics, these policies were aimed solely at optimization of timber production for export, primarily to the U.S., in a staples economy.[1] Consequently, these policies emphasize timber harvesting rights and long-term security for private investors through renewable leases. Within Canada, 80 percent of annual fibre allocations are associated with large-scale processing facilities that require minimum fibre flows to be economically viable, and 100 percent of the annual cut is designated to come from specific forest areas (Haley and Nelson 2007). Consequently, it has become quite difficult to adjust policies, reconfigure mill operations, and redirect fibre flows without affecting specific mills, towns, and forests. Under what has remained a timber- and volume-focused industry for more than 100 years, communities have had few avenues for developing and implementing the sorts of changes many believe are necessary to realize sustainability. The dominant tenure systems throughout Canada do not encourage private companies to engage the spectrum of cultural and environmental values increasingly demanded by communities (Haley and Nelson 2007; Luckert et al. 2011).

For various groups, and at different moments in Canada's history, conventional approaches to forest policy making and practice have created unacceptable levels of ecological degradation, economic instability, and social conflict. Negative outcomes have led some to search for alternatives to the long-time prevailing model. A shifting constellation of public values and beliefs regarding forests, and evolving understanding of the relationships between communities and forest ecosystems, have brought about gradual change in expectations regarding what constitutes acceptable use and development of public forests (Beckley 2003). Public demands to adapt forest policies thus reflect changes at the local and societal levels as communities, the forest sector, and public-at-large adjust to changes induced by micro and macro factors.

A mix of different forces has created space for new issues and actors in the forest governance arena, and community forestry has benefited from several parallel initiatives and changes. First, alongside changing public attitudes concerning forests and forest use, of major significance is the Indigenous empowerment agenda (Wyatt 2008). Shifting public attitudes, supported by court decisions upholding Indigenous rights and endorsement of the concept of free, prior, and informed consent (FPIC) in the United Nations Declaration on the Rights of Indigenous Peoples (Gallagher 2012; Coates and Newman 2014), are building support for increased Indigenous involvement in forest governance and development. This has helped Indigenous communities, as well as their partners and neighbouring communities, to become more involved in management.

Second, environmental groups, both international and domestic, have successfully lobbied for stronger environmental requirements (Luckert et al. 2011). The rise of third-party forest certification has helped to bring other perspectives into forest management decisions. Third, at the same time, the companies and governments that once formed a closed policy relationship and directed the Canadian forest sector, labelled the "business and government nexus" by Howlett and Rayner (2001), have weakened. Industry has been deeply affected by competitiveness issues, namely a decrease in both quantity and quality of wood supply, rising energy costs, an aging workforce and labour shortages, for example. Government forest departments have been subjected to streamlining or rationalization and, at least for the short term, decreased revenues from forestry. Amid the changing context for forest policy and practice in Canada, as in many parts of the world, "governments are no longer the most important source of decision making" (Armitage et al. 2012: 1).

Finally, given the economic instability, environmental impacts, and social unrest associated with resource dependence, communities have slowly come to demand more responsibility in forest management. From a community-based perspective, increased control over resource opportunities is seen as necessary to develop local capacities, curb unwanted demographic changes (i.e., whether to reverse youth out-migration or amenity migrant inflow), preserve ways of life, and, ultimately, to limit alteration of the internal structure of communities in forest regions undergoing rural transition (see Bullock et al. 2009). All of the above factors support the notion that forest management and planning in Canada are gradually becoming more "inclusive and open" (McGurk et al. 2006: 809). Understanding such processes and micro factors of change are the subject of a growing body of research, discussed in the section below.

Evolving Community Forest Research and Understanding

The need and opportunity for community forestry research, information, and infrastructure to support dissemination are both at an all-time high. There are now approximately 119 community forests in Canada, most of which were formed during the past two decades (Teitelbaum 2016). Since the 1920s, different provincial governments have introduced programs to enable the formation of community forests (e.g., Ontario in 1922, 1946, 1991; Quebec in 1993, 1997, 2001, 2004; British Columbia in 1998; Nova Scotia in 2013), which in turn has boosted demand for information to advise decision makers. At the same time, new arrangements, such as Indigenous and Indigenous-municipal partnerships, are emerging and causing community leaders and policy makers to look for examples, information, and partners from other regions and even other sectors.

Growth in community forestry research is evident when considering the increasing number of researchers involved with communities and contributing to the literature (Bullock and Lawler 2015). Likewise there has been steady growth in the number of journal articles produced and the range of themes addressed. It has been an interdisciplinary approach, integrating both social and natural sciences, that has advanced community forestry research. Whereas applied and evaluative research are now offering valuable insights on practice and policy outcomes (e.g., Teitelbaum 2016; Furness et al. 2015; Pinkerton and Benner 2013; Ambus and Hoberg 2011; Teitelbaum and Bullock 2012), earlier studies focused on conceptualizing community forestry and optimal conditions for its implementation (Dunster 1989; Duinker et al. 1991) or describing the operations and experiences of organizations (Allen and Frank 1994; Teitelbaum et al. 2006; Bullock et al. 2009; McIlveen and Bradshaw 2009). Research expansion has been driven by universities; however, non-governmental initiatives and communities have also contributed valued information. As recent examples, several guidebooks have been produced for practical reference, including the British Columbia Community Forest Association–sponsored *Community Forestry Guidebook II: Effective Governance and Forest Management* (Mulkey and Day 2012) and the *Guide to Community Forests in Nova Scotia* (MacLellan 2013). While the corpus of information available to community forest decision makers has expanded, a means to coordinate and disseminate this information in an interactive fashion has not yet emerged.

With respect to geographical representation, British Columbia and Ontario have remained focal regions for community forestry research, in terms of the number of both active researchers and research sites. The northern Territories, the Prairies, and to some extent the Maritimes and Quebec are less represented in community forestry research (Bullock and Lawler 2015). This is perhaps not surprising given that the forest sectors in these regions (save Quebec) are smaller in comparison to other parts of Canada (see NRCAN 2014). Remarking on the great diversity across community forests in general, McCarthy (2005: 995) points out that the emergence of community forestry "reflects changes in dominant models of forest management around the globe and the rapid diffusion of 'best practices' through networks linking forestry professionals, governments, conservation nongovernmental organizations (NGOs), and activists." Efforts to comprehensively document Canadian experiences with community forestry point to new opportunities for research engagement and collaboration involving a broad array of interested groups and disciplines and, as discussed below, there are several supporting initiatives now already underway across the country.

Advancing Community Forests through Advocacy

Communities and their partners have advocated for community forestry for decades in Canada. Notably, advocacy by First Nations and Metis for greater control of forests has been long-standing due to their marginalization from conventional forestry decision making and benefits. Many initiatives that have arisen from these Indigenous advocacy efforts are unique but related to community forestry because they share the core principles outlined in Table 0.1. These include Crown forest licences held by First Nation management companies, such as the Whitefeather Forest in Ontario's far north and Northwest Communities Forest Products Ltd. in Saskatchewan. Furthermore, some First Nation organizations are pressing for what they self-define as community forests in jurisdictions where they have not yet been enabled. These include Nishnawbe Aski Nation (NAN 2015) and Northeast Superior Regional Chiefs' Forum in northern Ontario (Lachance, Chapter 5), and the Ross River Dene Council in the Yukon.

Parallel to the widespread Indigenous push for decision-making power over forests, community forest advocacy has materialized both where community forestry has been enabled by legislation and where this has not occurred to date. Non-governmental organizations, including environmental NGOs (ENGOs) and model forests in some cases as well as academics, have played a key role in raising debates about the use of public forests and in mobilizing citizens and stakeholder groups to press for alternatives to the industrial forestry paradigm.

This advocacy has been mostly independent of any coordinated national support given that each province has jurisdiction over its own Crown forests (Bullock and Hanna 2012).

Regionally, Quebec has a long history of grassroots advocacy (>100 years) based on an early vision for community forests arising from a strong connection between forests and people and collective concerns about sustainable forest management (Nadeau and Teitelbaum 2016). The resulting collective action led to initial forms of local and regional forest governance models, though they did not achieve Crown tenure rights. In the 1990s, strong advocacy efforts by Solidarité rurale du Québec, a coalition of rural organizations, resulted in the establishment of the first community forests on Crown lands. A subsequent social movement for Inhabited Forests—the occupancy and use of forest lands by rural communities to ensure ecosystem sustainability and a place to live—promoted by Opérations Dignités was successful in achieving a number of short-lived pilots but no legal framework to enable this model. The culture of advocacy for community forests in Quebec has most recently manifested in a push for socio-economic development in rural communities and First Nations through Local Forests (Forêts de Proximité, see Laplante and Provost 2010), an approach that is now under development through several pilot projects.

The Algonquin Forest Authority was created on Crown land in the mid-1970s as a provincial response to conflict between the Algonquin Wildlands League (AWL) (currently Canadian Parks and Wilderness Society–Wildlands League) and loggers in Algonquin Provincial Park (Killan and Warecki 1992). Although at the time AWL did not intentionally advocate for community forestry, this ENGO was nevertheless instrumental in raising public awareness about concerns over industrial forestry being sanctioned in an iconic wilderness park and was thus a key contributor to the emergence of the community forest concept. Community forestry on Ontario Crown lands advanced incrementally from that point forward until the forest sector downturn, which reached a crisis level by 2005, catalyzed the formation of the Northern Ontario Sustainable Communities Partnership (NOSCP) in 2006. This multi-party NGO has advocated since its inception for community forests to promote resilience of communities and their local forests in northern Ontario (NOSCP 2010, 2011). NOSCP developed and released for endorsement a community forestry charter (see NOSCP 2007) soon after its inception, hosted/co-hosted several workshops and a conference on community forestry, and has been an active participant in the provincial forest tenure reform process that began in 2009 (see Palmer and Smith, Chapter 2).

In some regions, despite growing support, community forestry has gained less traction. For example, New Brunswick does not have a provincial community program. However it is a predominantly rural region where residents have strong connections to local forests and advocacy for community forests on public lands has been ongoing for the past two decades by communities, academics, and ENGOs. The Conservation Council of New Brunswick (CCNB) has taken a leading role in promoting community forests since the early 1990s and is a key ENGO in the New Brunswick Community Forestry Alliance, which includes a wide range of stakeholders. The alliance has developed a charter of shared principles for community forestry in New Brunswick (NBCFC n.d.) and provides ongoing support and advocacy for community forest proposals on Crown lands.

In other Atlantic provinces, including Nova Scotia (see MacLellan and Duinker, Chapter 6) and Newfoundland (Kelly and Carson 2016), advocacy for community forests is a much more recent phenomenon. It has emerged only in the past several years, from citizens, local governments, NGOs, academics, and economic development organizations, as a result of the forest sector downturn that saw the collapse of the mainstream forest industry in these regions.

In British Columbia, public awareness about the negative impacts of industrial logging in coastal temperate rainforest ecosystem was first raised by ENGOs and First Nations in the 1970s and 1980s. High-profile protests (e.g., Clayoquot Sound in 1984, Haida Gwaii in 1995) culminated in the "War in the Woods" by the 1990s, when forest industry came under attack by labour, environmentalists, communities, and First Nations (Wilson 1998), setting the stage for the ensuing public discourse about the need for alternative approaches. The Tin Wis Coalition of unlikely partners (ENGOs, trade unions, small businesses, and First Nations) was subsequently formed to promote community forests (Pinkerton 1993) in the midst of municipal conventions and conferences to debate the approach. Academics also played a key advocacy role during this period (e.g., Pinkerton 1993; Curran and M'Gonigle 1999; M'Gonigle et al. 2001). Shortly after the implementation of an initial group of community forests under the enabling legislation that was created in 1998, the BC Community Forest Association was created in 2002 to support established and proposed community forests (Gunter and Mulkey, Chapter 8). This is a unique NGO in Canada, a grassroots and inclusive association of member community forests that provides ongoing support and advocacy on their behalf, in order to improve practice and expand the number of initiatives.

Given the range of community forestry advocacy activities and organizations throughout the country, there is a need for knowledge sharing and coordination in the form of a national network to help strengthen localized efforts. This is the case for jurisdictions where community forestry has not been endorsed through legislation and policy, but also where it has, in order that the approach can expand beyond its current minimal level.

Réseau Canadien de l'Environnement (RCEN), or the Canadian Environmental Network, an umbrella group for ENGOs in Canada, held an international community forest workshop in 2007 supported by a variety of Canadian and international donors, during which community forest practitioners and promoters from throughout Canada were brought together with researchers, advocates, and practitioners from around the world. During this event, which produced a *Global Call to Action for Community Forestry in the 21st Century* (RCEN 2007), a seed was planted for the establishment of a pan-Canadian community forest network.

This concept came to fruition when in 2013 a number of organizations from throughout Canada that have actively advocated for community forestry (universities, NGOs, RCEN, Indigenous organizations, municipalities), with support from various funding partners including the Social Sciences and Humanities Research Council of Canada, partnered to co-host the country's inaugural community forestry conference in Sault Ste. Marie, Ontario. Key outcomes of this event were the establishment of a new national network, Community Forests Canada (CFC) (Palmer et al. 2013) and the idea for this volume. An author workshop that contributed to the book's production took place the following year in conjunction with a community forestry symposium to further expand the fledgling network (Bullock and Lawler 2014).

The Structure of this Book

Contributors to this volume represent a national community forest network that has emerged in the form of Community Forests Canada. They include researchers, practitioners, Indigenous representatives, government officials, local advocates, and students who are actively engaged in sharing experiences, resources, and tools of significance to forest resource communities, policy makers, and industry. Including research papers, professional essays, and vignettes, this volume addresses the emergence of community forests and initiatives to promote them throughout Canada and internationally. This edited collection is an attempt to bridge practice, research, and advocacy. As such it provides a timely resource to help advance community forestry as a process for collaborative decision making that links natural resource and community planning toward environmental stewardship, socio-economic development, and cultural autonomy.

The first major section of this book is dedicated to research that addresses the advancement of community forest initiatives through *evolving forest governance and collaborative networking*. In many jurisdictions institutional reforms are necessary to create supportive policies, redress power imbalances, and build relationships among Indigenous and non-Indigenous actors. These chapters describe the creative and often provocative changes and arrangements that have been attempted to enable community forestry. In **Chapter 1**, Ryan Bullock, Sara Teitelbaum, and Julia Lawler provide an overview of the diverse organizational arrangements that are typically used across the community forestry landscape in Canada. They apply the lenses of property and ownership regimes, governance rights, management objectives, and the legal entities to help make sense of the diversity in practical experiences. **Chapter 2**, by Lynn Palmer and M.A. (Peggy) Smith, provides an in-depth look at how transformative community organizing practice can contribute to advancing community forestry at both a regional and national level. The Northern Ontario Sustainable Communities Partnership (NOSCP) is a clear example of how inclusive peer networks, based on education and advocacy, can shift power dynamics within communities and regions, to elevate the collective voice of those who have been both disenfranchised and disempowered by current forest tenure policy. In **Chapter 3**, Stephen Harvey points to many policy gaps between forests and communities. He discusses the evolution of northern development and forestry policy and planning exercises as they relate to Indigenous and non-Indigenous communities embedded within northern Ontario's forests. Harvey's review highlights key changes during the recent period of forest policy reforms and implications for community involvement in forestry. Drawing on thirty years of direct experience with community forestry policy in northern Ontario, he offers professional recommendations that, if addressed, could help to bridge community forestry practice and policy gaps.

One of the key themes of this book is the inherent right of Indigenous peoples to participate in decision making regarding the future of forests in their territories. Chapters 4 and 5 offer a researcher's perspective and a practitioner's perspective on attempts to build collaborations in forestry management. In **Chapter 4**, Giuliana Casimirri and Shashi Kant explore cross-cultural collaboration that reveals some of the factors which either hinder or promote such practices and establishes that recognizing differing worldviews among collaborators is crucial. The chapter also demonstrates that building equitable decision-making structures is a multi-dimensional process requiring much more than simply providing a "seat at the table." In **Chapter 5**, Colin Lachance provides a practitioner's perspective on the almost decade-long implementation efforts of the Northeast Superior Regional Chiefs' Forum to advance a common forestry agenda by strengthening relationships among Indigenous communities and municipalities in a region of northern Ontario. Lachance's focus emphasizes the benefits of basing such relationships on traditional Indigenous knowledge and concludes that collaboration between First Nations and municipal governments needs to move beyond political advocacy and the forestry sector to service provision and other sectors, such as energy and mining.

In **Chapter 6**, Kris MacLellan and Peter Duinker examine the Nova Scotia government's recent policy interest in community forestry while also raising a caution that local governance of forests does not always lead to ideal conservation practice and that both policy makers and practitioners need to be mindful of the dangers of acting on unsupported assumptions. Their arguments support the need for research closely tied to policy and practice to ensure that critical questions are addressed in any process of forest management. They identify a five-step process of rigorous research of a diversity of models, based on broad consultations and community engagement, to establish a way forward to conserving and expanding the province's forests. In **Chapter 7**, Tracy Glynn describes the emergence of community capacity and infrastructure to support development of a community forest in the Upper Miramichi area of New Brunswick. Starting with the amalgamation of a number of small communities into one larger regional municipality to strengthen the political clout of the region, Glynn traces a collaborative process of undertaking a strong public relations campaign, including significant work "in the trenches" to build relationships and support across the region and an alliance of both Indigenous and non-Indigenous communities.

The chapters in the second major section make a case for how community forestry and forms of involvement can *build community capacity and promote enterprise and diversification*. Jennifer Gunter and Susan Mulkey begin this section in **Chapter 8**, with a review of the more than fifteen years of community forest tenure in BC and identification of key components of success. Political support at all jurisdictional levels, skills in organizational governance matched with appropriate supporting resources for proper management, and a network of practitioners whose needs are attended to and where knowledge is shared are all crucial success factors. BC's recent introduction of forest tenure for First Nations, although limited to timber rights, once again positions this province as a leader. In **Chapter 9**, researchers Felicitas Egunyu and Maureen G. Reed explore one of the common themes of the book—social learning through community forest governance—by examining the Harrop-Procter Community Forest in British Columbia as a case study. Here, social learning, that is, learning that moves beyond the individual to communities of practice, was demonstrated in a variety of different fields, including knowledge gained about: community forestry and community forestry management practices; co-operatives and administration/participation in a co-operative; working with limited resources and making the necessary tough decisions; and fostering innovation, though innovation may also come with risks, including risking (ironically) continued social learning. The researchers conclude that community forests can provide a rich site for social learning and reinforce other research findings that the local context encourages a broad diversity of management and business models.

Chapter 10 identifies yet another site for social learning and adds a further dimension to the Harrop-Procter case by examining how climate change impacts ecosystem resilience and the need for community forests to adapt to this threat. Practitioner Erik Leslie outlines the serious consequences of fighting forest fires based on assumptions that do not take into account the changing ecosystem conditions resulting from climate change, identifying steps that the Harrop-Procter co-operative has taken to mitigate these evolving issues. In **Chapter 11**, authors Brenda Murphy, Annette Chretien, and Grant Morin provide insight into the Indigenous–settler relationship in forest use and management, identified throughout this volume as crucial to community forest success, through an examination of the value-chain of maple syrup production. They argue that non-timber forest products should be incorporated into community forestry, which itself can be a bridge between scientific and Indigenous knowledges and can be a site to explore shared values and creative innovation. Their chapter raises many questions about how forests are valued and who is determining their value, as they explore in depth the differing worldviews between Indigenous and non-Indigenous maple syrup producers.

Chapter 12 concludes this second section of the volume with an examination of the economics of community forests. David Robinson suggests that there are several reasons why community-managed forests should provide greater economic benefit than those that are industrially managed. He points out that community forests are able to benefit from the same timber harvesting (if they choose) as industrial forests, but they also have access to many other benefits: the social learning, for example, that is outlined by other authors in this volume; the potential to allocate some or all of the forest to the harvesting of non-timber forest products; and the benefits that accrue, including innovation and creativity, from diverse perspectives, worldviews, and knowledges being shared. Robinson also argues that through local, public governance, costs associated with regulation of industry, necessary in the currently dominant industrial tenure system, can be avoided. In sum, community forestry is "an expansion of possibilities, not a set of restrictions." The concluding chapter synthesizes emerging themes, understanding, and practical actions for helping to bridge practice, research, and advocacy for communities and their forests.

Notes

1 Staples are raw or unfinished bulk commodity products sold in export markets with minimal amounts of local processing, as is the case for most Canadian forest products (Howlett and Brownsey 2008).

References

Alemagi, D. 2010. "A Comparative Assessment of Community Forest Models in Cameroon and British Columbia, Canada." *Land Use Policy* 27 (3): 928–36.

Allan, K., and D. Frank. 1994. "Community Forests in British Columbia – Models that Work." *Forestry Chronicle* 70 (6): 721–24.

Ambus, L., and G. Hoberg. 2011. "The Evolution of Devolution: A Critical Analysis of the Community Forest Agreement in British Columbia." *Society and Natural Resources* 24 (9): 933–50.

Apsey, M., D. Laishlef, V. Nordin, and P. Gilbert. 2000. "The Perpetual Forest: Using Lessons from the Past to Sustain Canada's Forests in the Future." *Forestry Chronicle* 76 (1): 29–53.

Armitage, D., R. de Loë, and R. Plummer. 2012. "Environmental Governance and its Implications for Conservation Practice." *Conservation Letters* 5: 245–55. doi: 10.1111/j.1755-263X.2012.00238.x.

Auden, A. 1944. "Nipigon Forest Village." *Forestry Chronicle* 20 (4): 209–61.

Barkley, B., M. Patry, P. Story, and S. Virc. 1997. "The Eastern Ontario Model Forest: Acting Locally, Connecting Globally." *Forestry Chronicle* 73 (6): 723–26.

Beckley, T. 1998. "Moving Toward Consensus-based Forest Management: A Comparison of Industrial, Co-managed, Community and Small Private Forests in Canada." *Forestry Chronicle* 74 (5): 736–44.

———. 2003. "Forests, Paradigms, and Policies Through Ten Centuries." In *Two Paths Toward Sustainable Forests: Public Values in Canada and the United States,* edited by B. A. Shindler, T. Beckley, and C. Finley, 18–34. Corvallis: Oregon State University Press.

Belsky, J. 2008. "Creating Community Forests." In *Forest Community Connections: Implications for Research, Management, and Governance*, edited by E. Donoghue and V. Sturtevant, 219–42. Washington, DC: RFF Press Book.

Benner, J., K. Lertzman, and E. Pinkerton. 2014. "Social Contracts and Community Forestry: How Can We Design Forest Policies and Tenure Arrangements to Generate Local Benefits?" *Canadian Journal of Forest Research* 44: 903–13.

Berkes, F. 2010. "Devolution of Environment and Resources Governance: Trends and Future." *Environmental Conservation* 37 (4): 489–500.

Bixler, R.P. 2014. "From Community Forest Management to Polycentric Governance: Assessing Evidence from the Bottom Up." *Society and Natural Resources* 27 (2): 155–69.

Booth, A.L. 1998. "Putting 'Forestry' and 'Community' Into First Nations' Resource Management." *Forestry Chronicle* 74 (3): 347–52.

Bradshaw, B. 2003. "Questioning the Credibility and Capacity of Community-based Resource Management." *Canadian Geographer* 47 (2): 137–50.

Brubacher, D. 1998. "Aboriginal Forestry Joint Ventures: Elements of an Assessment Framework." *Forestry Chronicle* 74 (3): 353–58.

Bullock, R., and K. Hanna. 2012. *Community Forestry: Local Values, Conflict and Forest Governance*. Cambridge: Cambridge University Press.

Bullock, R., K. Hanna, and D.S. Slocombe. 2009. "Learning from Community Forestry Experience: Challenges and Lessons from British Columbia." *Forestry Chronicle* 85 (2): 293–304.

Bullock, R., C. Keskitalo, T. Vuojala-Magga, and E. Laszlo Ambjörnsson. 2015. Forestry Administrator Framings of Responses to Socio-Economic Disturbance: Examples from Northern Regions in Canada, Sweden and Finland. *Environment and Planning C: Government and Policy* 34 (5): 945–62.

Bullock, R., and J. Lawler. 2014. *Community Forests Canada: Bridging Practice, Research and Advocacy.* Workshop and Symposium Report, Centre for Forest Interdisciplinary Research and Department of Environmental Studies and Sciences, University of Winnipeg, Winnipeg, MB.

———. 2015. "Community Forestry Research in Canada: A Bibliometric Perspective." *Forest Policy and Economics* 59: 47–55.

Bullock, R., and M. G. Reed. 2016. "Towards an Integrated System of Communities and Forests in Canada." In *Community Forestry in Canada: Drawing Lessons from Policy and Practice,* edited by S. Teitelbaum. Vancouver: University of British Columbia Press.

Burton, P.J., C. Messier, G.F. Wetman, E.E. Prepas, W.L. Adamowicz, and R. Tittler. 2003. "The Current State of Boreal Forestry and the Drive for Change." In *Towards Sustainable Management of the Boreal Forest*, edited by P.J. Burton, C. Messier, D.W. Smith and W.L. Adamowicz, 1–40. Ottawa: NRC Research Press.

Cain, W.C. 1939. "Forest Management in Ontario." *Forestry Chronicle* 15 (1): 16–28.

Charnley, S., and M.R. Poe. 2007. "Community Forestry in Theory and Practice: Where Are We Now?" *Annual Review of Anthropology* 36: 301–36.

Coates, K., and D. Newman. 2014. "Trilhqot'in Ruling Brings Canada to The Table." *Globe and Mail.* http://www.theglobeandmail.com/globe-debate/tsilhqotin-brings-canada-to-the-table/article20521526/.

Curran, D., and M. M'Gonigle. 1999. "Aboriginal Forestry: Community Management as Opportunity and Imperative." *Osgoode Law Journal* 37 (4): 711–774.

Drushka, K. 2003. *Canada's Forests: A History*. Montreal: McGill-Queen's University Press.

Duinker, P., P. Matakala, F. Chege, and L. Bouthillier. 1994. "Community Forests in Canada – An Overview." *Forestry Chronicle* 70 (6): 711–20.

Duinker, P., P. Matakala, and D. Zhang. 1991. "Community Forestry and its Implications for Northern Ontario." *Forestry Chronicle* 67 (2): 131–35.

Dunster, J. 1989. "Concepts Underlying a Community Forest." *Forest Planning Canada* 5 (6): 5–13.

Furness, E., H. Harshaw, and H. Nelson. 2015. "Community Forestry in British Columbia: Policy Progression and Public Participation." *Forest Policy and Economics* 58: 85–91. http://dx.doi.org/10.1016/j.forpol.2014.12.005.

Gallagher, B. 2012. *Resource Rulers: Fortune and Folly on Canada's Road to Resources*. Waterloo, ON: Bill Gallagher.

Grainger, S., E. Sherry, and G. Fondahl. 2006. "The John Prince Research Forest: Evolution of a Co-Management Partnership in Northern British Columbia." *Forestry Chronicle* 82 (4): 484–95.

Haley, D., and H. Nelson. 2007. "Has the Time Come to Rethink Canada's Crown Forest Tenure Systems?" *Forestry Chronicle* 83 (5): 630–41.

Hammond, H. 1990. "Community Control of Forests." *Forest Planning Canada* 6 (6): 43–46.

Hayter, R., and T. Barnes. 1997. "Troubles in the Rainforest: British Columbia's Forest Economy in Transition." In *Troubles in the Rainforest: British Columbia's Forest Economy in Transition*, edited by T. Barnes and R. Hayter, 1–11. Victoria, BC: Western Geographical Press.

Hickey, C., and H. Nelson. 2005. *Partnerships Between First Nations and the Forest Sector: A National Survey*. Sustainable Forest Management Network. https://era.library.ualberta.ca/files/t722hb08n#.WH0GsvkrK70.

Hessing, M., M. Howlett, and T. Summerville. 2005. *Canadian Natural Resource and Environmental Policy: Political Economy and Public Policy*. Vancouver: University of British Columbia Press.

Howlett, M., and K. Brownsey. 2008. "Introduction: Toward a Post-staples State?" In *Canada's Resource Economy in Transition: The Past, Present, and Future of Canadian Staples Industries*, edited by M. Howlett and K. Brownsey, 3–15. Toronto: Edmond Montgomery.

Howlett, M., and J. Rayner. 2001. "The Business and Government Nexus: Principal Elements and Dynamics of the Canadian Forest Regime." In *Canadian Forest Policy: Adapting to Change*, edited by M. Howlett, 23–64. Toronto: University of Toronto Press.

Kelly, E.C., and S. Carson. 2016. "The Roots of Community Forestry: Subsistence and Regional Development in Newfoundland." In *Community Forestry in Canada: Drawing Lessons from Policy and Practice*, edited by S. Teitelbaum. Vancouver: University of British Columbia Press.

Killan, G., and G. Warecki. 1992. "The Algonquin Wildlands League and the Emergence of Environmental Politics in Ontario, 1965–1974." *Environmental History Review* 16 (4): 1–27.

Krogman, N., and T. Beckley. 2002. "Corporate 'Bail-Outs' and Local 'Buyouts': Pathways to Community Forestry?" *Society and Natural Resources* 15 (2): 109–27.

Kube, M. 2012. "Community-based Forest Management." *Forestry Chronicle* 88 (2): 101–3.

Laplante, R., and C. Provost. 2010. "La cas de Champneuf et l'émergence de la notion de la forêt de proximité [online]." Montreal: Institut de recherche en économie contemporaine. http://www.irec.net/upload/File/Champneuf_fevrier_2010.pdf.

Lucas, R. 1971. *Minetown, Milltown, Railtown: Life in Canadian Communities of Single Industry.* Toronto: University of Toronto Press.

Luckert, M., D. Haley, and G. Hoberg. 2011. *Policies for Sustainably Managing Canada's Forests.* Vancouver: University of British Columbia Press.

Mabee, H., and G. Hoberg. 2006. "Equal Partners? Assessing Co-Management of Forest Resources in Clayoquot Sound." *Society and Natural Resources 19: 875–88.*

MacDonald, J.A. 1935. "A Plan for Reforestation Relief Works Projects in Southern Ontario." *Forestry Chronicle* 11 (2): 133–53.

MacLellan, K. 2013. *Guide to Community Forests in Nova Scotia.* Nova Forest Alliance, Stewiacke, NS. http://www.peiforests.ca/uploads/nfa/documents/Community_Forests_Guidebook_-_NFA_2014FINAL.pdf.

Marshall, F. 1986. "Community Forest Licences, and Other Thoughts on Wise Forest Management." *Forest Planning Canada* 2 (6): 8–11.

McCarthy, J. 2005. "Devolution in the Woods: Community Forestry as Hybrid Neoliberalism." *Environment and Planning A* 37 (6): 995–1014.

McGurk, B., A.J. Sinclair, and A. Diduck. 2006. "An Assessment of Stakeholder Advisory Committees in Forest Management: Case Studies from Manitoba, Canada." *Society and Natural Resources.* 19: 809–826.

McIlveen, K., and B. Bradshaw. 2009. "Community Forestry in British Columbia, Canada: The Role of Local Community Support and Participation." *Local Environment* 14 (2): 193–205.

M'Gonigle, M., and J. Dempsey. 2003. "Ecological Innovation in an Age of Bureaucratic Closure: The Case of the Global Forest." *Studies in Political Economy* 70: 97–124.

M'Gonigle, M., B. Egan, and L. Ambus. 2001. *The Community Ecosystem Trust: A New Model for Developing Sustainability.* Victoria, BC: POLIS Project on Ecological Governance, University of Victoria.

Mulkey, S., and J.K. Day, eds. 2012. *The Community Forestry Guidebook II: Effective Governance and Forest Management.* FORREX Series Report No. 30. Kamloops: FORREX Forum for Research and Extension in Natural Resources and Kaslo: British Columbia Community Forest Association. http://bccfa.ca/wp-content/uploads/2013/03/FS30_web-proof.pdf.

Nadeau, S., and S. Teitelbaum. 2016. "Community Forestry in Quebec: A Search for Alternative Forest Governance Models." In *Community Forestry in Canada: Drawing Lessons from Policy and Practice*, edited by S. Teitelbaum. Vancouver: University of British Columbia Press.

[NAN] Nihnawbe Aski Nation. 2015. *Ontario's Forest Tenure Modernization Act.* http://www.nan.on.ca/article/ontarios-forest-tenure-modernization-act--465.asp.

[NBCFC] New Brunswick Community Forestry Charter. n.d. file:///C:/Users/Ly/Downloads/charter.pdf.

Nelles, H. 2005. *The Politics of Development: Forests, Mines and Hydro-Electric Power in Ontario, 1849–1941,* 2nd ed.. Montreal: McGill-Queen's University Press.

[NOSCP] Northern Ontario Sustainable Communities Partnership. 2007. The Charter. Accessed 19 October 2015. http://noscp.ca/?page_id=37/.

———. 2010. "Response to Ontario's Proposed Framework to Modernize Ontario's Forest Tenure and Pricing System." Submitted to Ontario Ministry of Northern Development, Mines and Forestry, 18 May. http://noscp.ca/wp-content/uploads/2012/09/NOSCP_Charter_Response-to-tenure-consultation_18may10v2.pdf.

———. 2011. "Ontario's Forest Tenure Modernization Act: A Timid Beginning with Tons of Potential." Press Release, 20 May. http://noscp.ca/wp-content/uploads/2012/09/NOSCP_Press_Release_19may111.pdf.

[NRCAN] Natural Resources Canada. 2014. *The State of Canada's Forests – Annual Report 2014.* http://cfs.nrcan.gc.ca/pubwarehouse/pdfs/35713.pdf.

———. 2016. *How Does the Forest Industry Contribute to the Economy?* Accessed 21 November 2016. http://www.nrcan.gc.ca/forests/report/economy/16517.

Palmer, L., P. Smith, and R. Bullock. 2013. "Community Forests Canada: A New National Network." *Forestry Chronicle* 89 (2): 133–34.

Parkins, J.R., S. Nadeau, L. Hunt, A.J. Sinclair, M.G. Reed, and S. Wallace. 2006. *Public Participation in Forest Management: Results from a National Survey of Advisory Committees.* Information Report NOR-X-409. Edmonton: Natural Resources Canada, Canadian Forest Service, Northern Forestry Centre. https://cfs.nrcan.gc.ca/publications?id=26570.

Parsons, R., and G. Prest. 2003. "Aboriginal Forestry in Canada." *Forestry Chronicle* 79 (4): 779–84.

Pinkerton, E. 1993. "Co-management Efforts as Social Movements: The Tin Wis Coalition and the Drive for Forest Practices Legislation in British Columbia." *Alternatives* 19 (3): 33–38.

Pinkerton, E., and J. Benner. 2013. "Small Sawmills Persevere While the Majors Close: Evaluating Resilience and Desirable Timber Allocation in British Columbia, Canada." *Ecology and Society* 18 (2): 34.

[RCEN] Réseau Canadien de l'Environnement. 2007. *Global Call to Action for Community Forestry in the 21st Century.*

Ribot, J. 2002. *Democratic Decentralization of Natural Resources: Institutionalizing Popular Participation.* Washington, DC: World Resources Institute.

———. 2004. *Waiting for Democracy: The Politics of Choice in Natural Resource Decentralizations.* Washington, DC: World Resources Institute.

Robson, M. 2014. "Relative Influence of Contextual Factors on Deliberation and Development of Cooperation in Community-Based Forest Management in Ontario, Canada." *Canadian Journal of Forest Research* 44 (1): 64–70.

Ross, M. 1995. *Forest Management in Canada.* Calgary: Canadian Institute of Resources Law.

Roy, M.A. 1989. "Guided Change Through Community Forestry: A Case Study in Forest Management Unit-17 – Newfoundland." *Forestry Chronicle* 65 (5): 344–47.

Teitelbaum, S. 2014. "Criteria and Indicators for The Assessment of Community Forestry Outcomes: A Comparative Analysis from Canada." *Journal of Environmental Management* 132: 257–67.

Teitelbaum, S., ed. 2016. *Community Forestry in Canada: Drawing Lessons from Policy and Practice.* Vancouver: University of British Columbia Press.

Teitelbaum, S., T. Beckley, and S. Nadeau. 2006. "A National Portrait of Community Forestry on Public Land in Canada." *Forestry Chronicle* 82 (3): 416–28.

Teitelbaum, S., and R. Bullock. 2012. "Are Community Forestry Principles at Work in Ontario's County, Municipal, and Conservation Authority Forests?" *Forestry Chronicle* 88 (06): 697–707.

Wilson, E. 1943. "Forestry in Post-war Rehabilitation." *Forestry Chronicle* 19 (1): 14–16.

Wilson, J. 1998. *Talk and Log: Wilderness Politics in British Columbia.* Vancouver: University of British Columbia Press.

Wyatt, S. 2008. "First Nations, Forest Lands, and 'Aboriginal Forestry' in Canada: From Exclusion to Co-management and Beyond." *Canadian Journal of Forest Research* 38: 171–80.

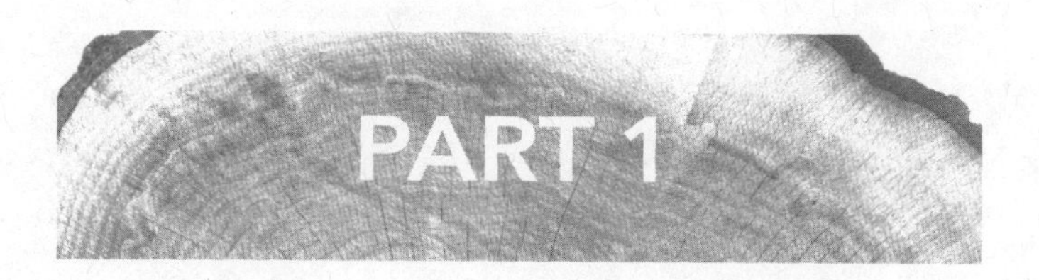

Advancing Forest Governance through Collaborative Networking

Characterizing Institutional Diversity in Canada's Community Forests

Ryan Bullock, Sara Teitelbaum, and Julia Lawler

Community forestry is a broad term that encompasses initiatives with different origins, institutional forms, and operational practices (Belsky 2008). This diversity is demonstrated in Canada, where community forestry has developed in concert with other forms of industrial and recreational forest development. However, community forests represent a small proportion of Canada's vast forest lands. The evolution of community forestry in most Canadian provinces is neither linear nor consistent, and many collectives and governments have experimented at one time or another with community-led models of forest management. A growing volume of research provides evidence for the spread and establishment of community forestry operations and concepts, most notably in British Columbia (Benner et al. 2014; Ambus and Hoberg 2011; Bullock et al. 2009; Pinkerton et al. 2008), Ontario (Teitelbaum and Bullock 2012; Bullock and Hanna 2012), and Quebec (Teitelbaum et al. 2006; Masse 2002), but also in Saskatchewan (Beckley and Korber 1996), New Brunswick (Betts 1997), Newfoundland (Roy 1989), and most recently, Nova Scotia (MacLellan and Duinker, Chapter 6).

Much like the varied geographic landscapes that serve as the setting for community forestry in Canada, the social contexts and underlying motivations for local control over forest resources are both diverse and nuanced (Bullock et al. 2009; Furness et al. 2015). Some community forest supporters are motivated by the need for economic development (Murphy, Chretien, and Morin, Chapter 11), and others are driven by a desire to better conserve forests and their resources amid environmental change (see Leslie, Chapter 10; also

Furness and Nelson 2012). Some local groups have sought control over nearby forests due to conflict in contentious areas (see Casimirri and Kant, Chapter 4, also Bullock and Hanna 2012). In some settings, local control of forest lands and resources is pivotal to restoring and securing culture, identity, and traditional stewardship roles (see Lachance, Chapter 5; Smith 2013; Passelac-Ross and Smith 2013; Bullock 2012). Many community forestry initiatives combine a number of these and other objectives suited to a particular place. Given these differences, it can be difficult to make generalizations about the underlying values of community forest organizations. Moreover, it can be challenging to identify exactly what distinguishes community forests from other management approaches, particularly given the recent push toward heightened public participation within provincial forest policy frameworks. The term "community forestry" may mean different things to different people, which is not inherently problematic. Nonetheless there is an opportunity to bring greater clarity to the discussion and start distinguishing the different forms community forestry often takes. Other researchers have begun the work of mapping out different conceptual dimensions of community forestry, namely Krogman and Beckley (2002) and later Belsky (2008), which we build upon in this chapter.

As Krogman and Beckley (2002: 111) point out, "part of the problem in assessing community forestry and community forests stems from the fact that these terms are used widely to describe a broad spectrum of conditions, experiments, and institutions." In the same paper Krogman and Beckley (2002) offer a bimodal typology that includes a simple continuum ranging from total control and benefits at one end to no control and benefits at the other, acknowledging that neither extreme could exist in Canada (or indeed the world). Krogman and Beckley's (2002) caveat points to the diversity that makes classifying community forests difficult, and the concomitant need for flexible approaches. Likewise, Belsky (2008), writing about community forestry in the U.S., takes a functional approach to analyzing local governance arrangements in order to distinguish Indigenous, municipal, and conservation organization–based community forests using attributes such as access and control, the nature of the community to be represented, organizational form, and ownership. In an earlier paper Beckley (1998) examined the locus and structure of decision making, management objectives, tenure type, as well as the scale of operations and knowledge base applied through industrial, co-managed, community, and small private woodlot forests. These attempts to classify community forestry by organizing key concepts and practical examples highlight that, at its root, community forestry emphasizes enabling involvement of non-conventional actors in forest management and development—involvement in decision making that sets the priorities, values, and actions to be imposed on forests, but also through receipt of various benefits, whether economic, social, or environmental (see Duinker et al. 1991). Enabling meaningful involvement requires governance infrastructure, such as institutions, organizational entities guided by a local vision and objectives. Thus, we examine resident and stakeholder interactions with forests under an array of local or regional arrangements intended to support local involvement through community forest management and development.

We present four main institutional perspectives prominently featured in relevant research on socio-economics and environmental resource governance that enable community forests to be described with more precision based on key distinguishing features. These include:

1. the type of land ownership and property regime(s) under which the community forest initiative operates;
2. the degree of governance rights it holds;
3. the core values and management objectives that drive the organization; and
4. the sorts of organizational models and legal entities used.

In providing a basis for descriptive classification, we are not concerned with delineating the boundaries of what community forestry is or is not. Rather, we map out key features that can enable community forestry practitioners, scholars and policy makers, and students to make better sense of various initiatives that are currently identified under the community forestry "umbrella." The discussion below outlines four main institutional perspectives and criteria by which differences can be classified. This provides a framework that we elaborate with examples drawn from relevant literature to help readers make sense of the diversity that presently characterizes community forestry in practice. Below, we discuss notions of property, rights, objectives, and organization as conditions that shape community forestry arrangements in Canada.

Land Ownership and Property Rights

Private, Public, and Common Rights to Forest Lands and Resources

While the notion of "property right" can be applied to material (e.g., land) and non-material items (e.g., ideas), in the context of community forestry the concept is usually used in the former sense to refer to a set of rules governing access to and control over lands and resources (see Waldron 2012). The notion that forests and their resources can be considered property which can be held by an individual or group raises questions regarding who the owners are and what exactly it is that is owned. As Luckert et al. (2011: 43) point out in the context of forest resources, "it is rights, never objects, that are owned. A property right permits its holder to use an asset in order to enjoy the resulting stream of benefits subject to certain conditions, obligations, and prohibitions." Forest resources can be subject to different property rights held by different individuals or groups through time or at the same time.

There are three main categories of property rights used to govern human behaviour regarding scarce and "contested resources" (Waldron 2012), which can be used to understand property right and ownership regimes in Canada's community forests: common, public/collective, and private. Each arrangement indicates something about how rights are held (i.e., individually or jointly via some representative legal entity), how rights can be exercised (i.e., who decides and benefits), and the relationship between the rights holder(s) and society (i.e., conditions of acceptable use) (Luckert et al. 2011; Waldron 2012).

(1) *Common property* is available for use by all or any members of a society where rules are established to ensure that one use or user does not stop another (e.g., a woodland that remains accessible for diverse use and users with no prescribed overriding use). In this instance, individuals decide how they use common resources.[1]

(2) *Collective* or *public property* involves permissible uses and users being determined by members of society together in light of social interests linked to the whole and through some form of collective decision-making process and structure. An example of this can be seen in Canada's provincial Crown forests, which in different places are subject to certain conditions of access and use set out by various representative groups in the public interest.

(3) *Private property*, on the other hand, confers a higher degree of autonomy whereby use and/or users are decided by individual land/resources holders—with little or no obligation of concern for broader society, where "private individual[s] or groups hold unfettered title to forest land and all its products" (Luckert et al. 2011: 54).

Given the preponderance of public or "Crown" land in Canada (90 percent) (Natural Resources Canada 2014), it is suitable to consider a public-private division of forest property. However, it should be recognized that public agencies representing citizens can and do hold private property, which may even be off limits to the general public. Examples are Ontario Conservation Authorities, which are corporations that manage municipal lands and also lands owned by the corporate body. Another example where the lines blur between different categories of property exists in First Nations with settled land claims in the Northwest Territories, where land titles are held by Indigenous organizations, in this case corporations, on behalf of band members (AANDC 2013). Alternately, where Crown forests are managed by various people or groups, both common and collective/public property rights can be held by different users, for different uses, and under different formal and informal arrangements, even on the same piece of land. Several, often competing, forest users associated with Crown forest resources have been identified, namely, companies that hold long-term timber harvesting rights and management responsibilities transferred from the provinces to them; individuals holding Aboriginal and Treaty rights; other licensees involved in resource taking such as hunters, fishers, tourism operators, loggers, and non-timber forest product harvesters; and finally, users with non-commercial or non-consumptive rights of access such as berry pickers, hikers, canoeists, and the like (Smith 1995).

Such arrangements can make it difficult to observe distinctions among different property rights and ownership regimes, as forest resources and human behaviours on a certain land base can be subject to different institutional arrangements that can confer varying rights simultaneously (discussed in the next section). For example, the recently formed community forest on Cortes Island in BC would have provincial Crown lands being managed collectively as public property insofar as forest management decisions were being made based on the desires of non-Indigenous residents and Klahoose First Nation band members; however, the arrangement would also uphold common property rights for traditional forest uses by First Nations. While the land base is comprised of provincial and federal lands, held in trust for the peoples of Canada, senior governments would assign certain management rights to island-level interests, in this case through a private legal entity. In contrast, there are examples of private forest properties held by individuals that operate somewhat like community forests in terms of creating access (albeit for a fee), producing economic benefits (jobs), and managing for environmental values (habitat) (e.g., Haliburton Forest and Wildlife Reserve). However, while such properties can be managed with broader public values and benefits in mind, there is no obligation for the private owner to do so, nor does decision making need to be inclusive, open, and representative. Community forests in Canada have been created on various lands using various property rights and ownership regimes (Table 1.1).

Table 1.1. Community forests under different sorts of land ownership and property rights.

LAND OWNERSHIP	EXAMPLES	BEGAN	PROPERTY RIGHTS	PROVINCE
Federal reserve	Eel Ground First Nation	1990	Common/Public	New Brunswick
Unceded reserve	Wikwemikong First Nation	1992	Common	Ontario
Provincial	Medway Forest Co-operative	2013	Public	Nova Scotia
County	Simcoe County Forest	1922	Public, Private	Ontario
Municipal	Cowichan Municipal Forest	1946	Public, Private	British Columbia
Private Company	Seigneurie Nicolas-Riou & Seigneurie du Lac-Metis	1994	Private	Quebec

Degree of Governance Rights

The question of the level of governance *authority* held by a community has been a central point of discussion and debate in the community forestry literature. A concept that is often referred to in this discussion is *devolution*. What has emerged from this discussion is strong consensus, arrived at from research in widely different settings, about the need for communities to acquire sufficient decision-making authority as a building block in implementing effective and meaningful community forests (Agrawal et al. 2008; Wollenberg et al. 2008; Shackleton et al. 2002). In reality, few communities acquire a complete transfer of authority, and thus "the central challenge is determining an appropriate distribution of authority for specific decisions between levels of governance—between the state and communities, and between different levels and agencies of the state" (Ambus and Hoberg 2011: 937).

In Canada, community forestry can include different institutional arrangements, some with greater and lesser levels of devolution. For example, several community forestry pilot projects, such as those implemented in Ontario and Quebec in the 1990s, placed insufficient emphasis on actually creating tenure arrangements to confer management authority to community entities, resulting in the creation of a number of initiatives which served an exclusively consultative role. Further, Ambus and Hoberg's (2011) analysis of the BC Community Forest Agreement found that provincial devolution efforts were not deep enough to instill systemic changes needed to achieve the community forest ideal. Ambus and Hoberg (2011) identified that the Province of British Columbia did not embrace reforms of legislation and regulations necessary to actually empower communities to become involved in strategic decision making, for example.

Likewise, despite recognizing the importance of operational decision making as a gateway to capacity building needed for elevated Indigenous control and benefits, Wyatt (2008) acknowledges that this level of decision making still represents a more limited form of involvement. Community forest managers remind us (see Leslie 2016) that certain rights are essential for local community forests organizations seeking to effectively manage provincial lands; in order to achieve forest management objectives, communities must have the authority to regulate land use patterns and to alter public landscapes and natural resources. Without these rights and authority, forest communities remain trapped in a dependency cycle wherein they must follow provincial directives that may not align with their objectives, needs, or regional context (Palmer et al. 2016).

Such considerations raise the question—at what point can we distinguish a community forest from a participatory process? This important question harkens back to Krogman and Beckley's (2002) idea that a sufficient level of "local control" must be acquired in order to characterize an initiative as a community forest. As they see it, decision-making power is central to defining a community forest as "an entity that has an explicit mandate and legal decision-making authority to manage a given land base for the benefit of the local community" (Krogman and Beckley 2002: 112).

For descriptive purposes we are not concerned with determining what level of authority is required for an initiative to be considered a community forest. However, we are interested in distinguishing initiatives according to how much power they hold. But how do we do this? A useful conceptual scheme is presented by Schlager and Ostrom (1992) in their description of common property and rights-based arrangements. The authors speak about "bundles of rights" which may be held by local users. First, they distinguish between *operational* and *collective-choice* rights. Operational-level rights, as their name implies, represent a more limited set of rights, such as the right to *access* (the right to enter a property) and the right of *withdrawal* (the right to extract a resource). Collective-choice rights imply a higher level of authority and are exemplified by the right to *management* (the right to regulate internal use patterns), *exclusion* (the right to determine who will have an access right), and *alienation* (the right to sell or lease the above collective-choice rights). Thus, collective choice rights are of greater strategic importance, because they allow users to define future operational-level rights. Some scholars argue that community-based management is only meaningful if communities are involved in rule making at multiple levels. According to Agrawal and Ostrom (2001: 492), "simply granting rights to undertake operational level actions is insufficient to justify claims of decentralization." This scheme is useful when thinking about community forestry, as it provides a platform upon which to analyze precisely what rights a community holds with regard to governance of local forests.

Providing an important complement to the work of Schlager and Ostrom (1992), Ambus and Hoberg (2011) developed an analytical framework which helps to characterize community forestry governance authority within a Canadian context. Their framework distinguishes between three levels of authority: strategic, tactical, and operational. Forestry management functions are then defined and classified according to these three levels. Their analysis of the British Columbia Community Forest Agreement Licence reveals a stronger tendency toward devolution of operational than tactical and strategic level rights. Significantly, we now have several useful frameworks at our disposal that can help us to distinguish the level of authority held by a community forest. This should enable identification of distinctions between those organizations which have higher and lower levels of governance authority.

Management Objectives

Community forests often have many demands placed upon them. They can exist in areas with vast and diverse populations or areas with small and seemingly more homogenous populations. Consequently, they are managed for diverse goals under unique governance approaches that have been specified by local contexts and histories (Vernon 2007). These conditions translate into very different motivations for local control of forest resources (Bullock et al. 2009). Community forests also face the reality that many have been left with a legacy of poor management (Booth 1998) or industrial use wherein some forests have been heavily logged and burned, in some cases several times (see Bullock et al. 2009; Vernon 2007; Betts 1997 for case examples). Management objectives flow from the local values to be represented and the local ecological conditions and provincial institutional setting that both provide and constrain management opportunities. In community forests, management objectives tend to fall under four general categories: economic, environmental, socio-cultural, and educational.

Economic Development

Job creation has long been a focus of community forestry in Canada. During the Second World War community forests were considered as a source of employment for troops returning from war, and large-scale work projects fit with the service role, but also work programs were needed to help with economic stimulation and conservation (see Wilson 1943). Beyond employment, community forestry is pursued to provide stability and renewal through local economic development and diversification (Markey et al. 2012; Krogman and Beckley 2002; Robinson et al. 2001). Realizing the full potential of value-added processing and employment by community forests still requires supportive regulatory and policy reform, however (Vernon 2007). Recent research also suggests that despite expansion of ongoing resource-sharing arrangements in provinces such as British Columbia, servicing, infrastructure, and capacity gaps need to be addressed to help support community forest–based development (Markey et al. 2012). According to Markey et al. (2012: 294–95), non-Indigenous and Indigenous communities involved in local forest–based development must work together and identify ways to "scale-up" in order to better market their products and get better prices, reduce transportation costs, and share skills and knowledge to strengthen competitiveness. This collaborative territorial model for regional economic development is part of the community forestry ideal, which indeed requires systemic change (see Bullock and Reed 2015). In addition to strategic relationships, there is a recognized need for research to develop value-added and non-timber forest products, markets, and supply chains (Teitelbaum and Bullock 2012; also Murphy, Chretien, and Morin, Chapter 11). There is new evidence that specialty mills associated with community forests can provide more than twice as many jobs per cubic metre as larger commodity mills (see Pinkerton and Benner 2013), which provides support for the role of community forests in local economic development. But the relationships between economic and community stability cannot be fully explained by employment figures alone. Namely, local entrepreneurs and community-minded leaders often possess a certain attachment and commitment to place that can anchor local businesses and labour. For example, research in forest communities in northern Ontario (Bullock 2012) and British Columbia (Pinkerton and Brenner 2013) has shown that owner-operated mills can exhibit a certain commitment to their communities and employees that can move them to keep operating during even the most difficult economic downturns. Economic development, as an objective of community forestry, is multi-faceted.

Environmental Conservation

Environmental degradation has always been a main driver or motivation in initiating community forests (Bullock and Hanna 2012). Consequently, achieving environmental sustainability is a key objective of many community forest initiatives. During the late nineteenth and early twentieth centuries, the town forest movement of eastern North America was a collective response by foresters, politicians, farmers, naturalists, municipal leaders, and many others, to remediate and protect against soil erosion and depletion of timber and water supplies, resulting from poor farming and land clearing practices (see McCullough 1995; OMNR 1986). For example, municipal and county forests in southern Ontario were heavily conservation focused to deal with blow sands on abandoned farms throughout southern Ontario (Dunkin 2008). Subsequently, the most recent upswell of interest in community forestry during the 1990s was driven by shifting public values regarding forests, protected areas, industrial logging practices, and the appropriate role for citizen groups and other actors in managing public forest ecosystems (Robinson et al. 2001; Charnley and Poe 2007).

An explicit assumption of community forestry is that communities will use more environmentally friendly forestry practices (Ambus and Hoberg 2011). First-hand experience with the negative impacts of poor land use practices typically motivate community forest managers to opt for smaller-scale and "ecologically-benign" approaches (Betts 1997: 247). The Eel Ground Community Forest, for example, focused its management practices on preserving wildlife habitat and biodiversity in order to restore the forest after a period of unrestricted cutting resulted in degradation (Betts 1997). Community forestry also represents an opportunity to mobilize traditional environmental knowledge to improve decision making for better environmental outcomes through sustainable forest management (Grainger et al. 2006). Community forests are usually managed for various environmental values, such as the conservation and protection of biodiversity, water and timber supplies, soil nutrients, endangered species habitat, natural hazards, and viewscapes, as well as environment-dependent purposes such as businesses that require certain environmental conditions be maintained (i.e., remoteness, views, trails, wildlife viewing opportunities, etc.) (Bullock 2007). In addition to remediating previously degraded lands, an overall environmental goal of community forestry is protection and maintenance of ecological services provided by well-functioning forest ecosystems (Charnley and Poe 2007).

Social Inclusion / Cultural Autonomy

Several community forests have been initiated to try to mitigate conflict through social inclusiveness in decision making and development, and the building of social capital among forest stakeholders (Bullock and Hanna 2012).

Numerous clashes have occurred involving different interests linked to Crown and park land, cottage tourism areas, First Nations traditional territories, and commercial forestry areas. Community forests have been created across Canada in what were at certain times some of the most conflicted landscapes (e.g., Cortes Island, Creston, Slocan Valley, BC; Northwest Communities, SK). In Ontario, several large-scale community forests and processes were the direct result of multi-party conflicts and Indigenous land use agendas (e.g., Westwind Forest Stewardship Inc., Algonquin Forest Authority, Wendaban Stewardship Authority; Whitefeather Forest Initiative). Local control of forest lands and resources is seen to enable social inclusion in development and also cultural determination.

There is a slowly emerging recognition of the importance of culture, community identity, and environmental citizenship related to community forestry movements (Booth and Muir 2013; Bullock et al. 2012; Bullock 2012; Booth and Skelton 2011; Kube 2012). Not surprisingly, this consideration is most commonly discussed with respect to Indigenous forestry and First Nations involvement in forest management. For some First Nations, recovering a stewardship role through community forestry is central to maintaining traditional ways of life—a role that is required to in turn address cultural responsibilities (Bullock 2010). This role emerges from a belief that there are links between the land and culture and that what affects one affects the other (Booth and Skelton 2011). Managing community forests for traditional values affirms culture as well as claims to territory. In this way Indigenous community forestry can be a vehicle for protecting and upholding marginalized cultures through actualizing First Nations "beliefs, values, norms, practices, or knowledge associated with forest landscapes" (Wyatt 2008: 176).

Research and Education

Supporting opportunities for research and education is another core objective for some community forests. As a supplement to economic, social, and environmental objectives, this can involve initiatives for public awareness, outdoor education, formal training and work experience, professional development, research, monitoring, and policy learning. In fact, provincial community forestry programs in BC, Ontario, and Nova Scotia have used the "pilot" label to signify an explicit intent to experiment with new models in order to generate policy lessons and best practices. Probably the most original example of a learning-oriented community forest initiative is the John Prince Research Forest, which was established in 1999 to be a "social and institutional experiment in co-management" (Grainger et al. 2006: 485). The project involved a 13,000-hectare research forest located on Crown land in the traditional territory of the Tl'azt'en Nation managed in co-operation with University of Northern British Columbia. The arrangement has been characterized as a community forest given its social objectives and management approach (Grainger et al. 2006). Integrating traditional and scientific knowledge and the transfer of lessons among First Nation communities looking to implement similar arrangements were important priorities (Booth 1998). Forest management contracts in Quebec, such as the Corporation de Gestion de la Forêt de l'Aigle, include demonstrations of silvicultural techniques and other public education activities. This organization also includes a research organization on its seven-member board alongside First Nations, wildlife, and recreation groups (Teitelbaum et al. 2006). In southern Ontario, public education programs involving school groups and wider citizen groups are key ongoing activities (Teitelbaum and Bullock 2012). Though some community forests explicitly incorporate education, research, and training into organizational, programming, and management objective language (e.g., Creston Valley Forest Corporation, see Teitelbaum 2014), in other cases the importance of community forestry as an educational forum is discovered as a process outcome rather than an explicit objective (Roy 1989).

Community forestry in Canada has also received support from academic researchers and institutions, which perhaps has placed emphasis on research and education as management objectives. Teitelbaum et al. (2006) observed that community forestry policy, programs, and practice in BC have been advanced by academic research programs, notably University of Victoria's POLIS Project and the Centre for Sustainable Community Development at Simon Fraser University, as well as research-oriented non-governmental organizations (e.g., Dogwood Initiative, British Columbia Community Forest Association). The same can be said for Ontario, where university researchers and NGOs have advocated for community forestry for more than two decades (e.g., formerly Lakehead University's Research Chair in Forest Policy and Management [see Duinker et al. 1994]; but also Northern Ontario Sustainable Communities Partnership; NORDIK Institute, Algoma University; INORD, Laurentian University [see Bullock and Hanna 2012; Bullock 2012]). The emergence of community forestry in Nova Scotia is getting similar support from Dalhousie University researchers and research-based organizations (i.e., Nova Forest Alliance, see MacLellan and Duinker, Chapter 6; Mersey Tobeatic Research Institute). In Manitoba, the provincial government and other stakeholders have engaged the Centre for Forest Interdisciplinary Research at the University of Winnipeg to explore options for tenure reform and local development that include community forest management approaches (see Lawler and Bullock 2017).

Organizational Models

Another way to distinguish between community forests is through the types of organizational models they adopt. Several organizational models have been adopted within the Canadian context, each with its own mechanisms by which to address governance issues (see, e.g., Gunter and Mulkey, Chapter 8, for specific examples regarding organizational models used by Community Forest Agreements in British Columbia). According to Ribot et al. (2006), important questions in this regard include:

(1) How can community interests best be represented?
(2) How can effective and mutually agreeable decision making be ensured?
(3) How can accountability, both upward to central governments and downward to the broader community, be maintained?

What follows is a description of common organizational models and examples, which have been portrayed both in the academic literature and the more hands-on publications developed for community forestry practitioners (see Gunter 2004; Tyler et al. 2007; Mulkey and Day 2012). Each has its particularities, and the choice of organizational model is generally a reflection of the needs, motivations, and underlying objectives and legal requirements of the project. However, each model has different implications in terms of forms of representation and transparency requirements.

Co-operative

A co-operative is "an organization owned by the members who use its services or are employed there" (CCA 2014). Co-operatives are guided by a series of common principles including voluntary adhesion, democratic member control, co-operation, autonomy, and concern for the community (CCA 2014). Co-operatives are legally incorporated businesses, based on a shareholder model. For example, the Harrop-Procter Community Co-operative has been operating since 1999 (see Leslie, Chapter 10; also Elias 2000). Recently a new type of co-operative has gained some traction in Quebec, called the solidarity co-operative. This model allows for the addition of a new type of member, namely that of a supporting member, which means that any person or organization which shares the co-operative's mission can join the organization.

First Nation Band Council

This is the political entity which represents a First Nation, as prescribed under the Indian Act. The band council is composed of an elected chief and band council. In Canada, band councils are elected for a two-year term. Many First Nations are also working toward other types of governance arrangements; however, from the perspective of community forestry attributions on Crown land, the band council appears to be the preferred organizational entity from the perspective of provincial governments. Band councils sometimes then assign responsibility for their community forests to band-owned development corporations. Such was the case with the community forest established on the Wikwemikong Unceded Indian Reserve (see Harvey and Hiller 1994). However, First Nations are also developing other types of community-based management arrangements, such as *co-management*, for example, whereby a board may be composed of both representatives from a First Nation and government representatives (see Grainger et al. 2006). Still other First Nations have reserve lands which they manage as community forests, though there is a need to develop the local capacity to do so (Wyatt 2008; Westman 2005).

Local Government

Local governments are comprised of different administrative entities, including municipalities, towns, counties, and regional municipalities, among others. These are government administrative entities that hold specific authority and responsibilities as set out by provincial and federal governments. They are based on an elected structure, usually composed of a mayor and a council. In Quebec, Ontario, and British Columbia, local governments have been one of the main recipients of Crown land community forest tenures. Examples include the Mission Municipal Forest in BC. In some cases, services and programs are funded by forest-based revenues via local government bodies. The earliest examples of local government-run community forests are reviewed by Allan and Frank (1994), Bullock and Hanna (2012), and Teitelbaum et al. (2006).

Hybrid Local Government

In Ontario, another model associated with community forestry is that of conservation authorities (see Teitelbaum and Bullock 2012). These are watershed-based corporate organizations which have a mandate to "establish and undertake, in the area over which it has jurisdiction, a program designed to further the conservation, restoration, development and management of natural resources other than gas, oil, coal and minerals" (Government of

Ontario 1990). Boards of directors are made up of neighbouring municipal elected officials or appointees thereof. Examples with active community forest programs include the Ganaraska Region, South Nation, and Grand River Conservation Authorities.

Partnership

These are legal entities that are made up of two or more persons (or organizations). The assets and liabilities are shared jointly by the partners. This can be a partnership between a First Nation and a municipality, as has been the case in British Columbia. For example, the Cortes Forestry General Partnership is a collaborative initiative of the Klahoose Forestry Limited Partnership No. 2 (representing Klahoose First Nation) and the Cortes Community Forest Co-operative (representing the non-Native community) (Mulkey and Day 2012).

Not-for-profit Organization or Society

These are groups that are organized around a specific activity with the stipulation that no personal financial gain is permitted. Not-for-profits may not distribute any profit or dividend to members; instead, profits must be reinvested in carrying out the organization's goals and objectives. Some community forests opt to reinvest any profits directly in forest management activities, while others redistribute profits through community projects and donations. Not-for-profits have a membership that holds the responsibility to elect or appoint a board of directors, and that administers the organization. Not-for-profits must establish bylaws and policies that steer the organization's activities. Requirements for public accountability are strong and include annual financial statements and annual general meetings (Mulkey and Day 2012).

Corporation

A corporation is a separate legal entity owned by its shareholders. It is a commonly adopted legal structure for community forests in Canada, and has been frequently used by municipalities, band councils, and other local government entities in order to create a separate business entity. Corporations are allowed to own assets, borrow money, and receive dividends from profitable operations (Tyler et al. 2007). Corporations are governed by a board of directors who are obliged to act in the interest of shareholders. Ownership is divided into shares, which can be held by any number of shareholders. Municipalities often opt to remain the sole shareholder of a corporation and can thus appoint the directors (usually a combination of elected officials and members of the public) (Mulkey and Day 2012). Corporations have fewer legal requirements for downward accountability than other models (reporting requirements, membership opportunities, for example); however, some have opted to go beyond legal requirements for transparency and public consultation by using open houses, field trips, and newsletters.

Summary

In this chapter we outline central concepts and examples for understanding and characterizing institutional diversity in Canada's community forests. By doing so we provide a heuristic to gain a more nuanced view of community forestry models—one that is more inclusive of its many configurations. In particular, this detailed framework calls attention to the kinds of property rights and land ownership arrangements community forests operate under; the level of authority and responsibilities they can hold; the core values and management objectives that drive the organizations; and the sorts of legal entities that are frequently utilized to establish community forests.

Applying this framework to undertake an extensive national classification of community forests will be the work of a future paper. However, we propose that each of the four main components outlined above can be used individually or holistically. Importantly, this work interprets and synthesizes current theory on institutions and community-based natural resource management to more completely capture the practical diversity of, and scholarly thought regarding, institutional arrangements under which community forestry plays out in Canada.

By characterizing the diversity of community forests, this chapter reveals several important opportunities for inquiry that could help fill gaps in understanding produced by the variety of contexts for, and approaches employed by, local forest management initiatives. In-depth and comparative assessments of property rights and related issues as they arise in community forests would help provide a more nuanced understanding of the complex interactions at work in forest landscapes and communities. There is more work to be done to better understand the intersection of different forest users and their rights (see Smith 1995), including community forest management organizations and the groups with which they collaborate and represent. Community forests are new players situated among a growing range and number of users that possess different rights, including: (a) companies with long-term harvesting rights and management responsibilities; (b) First Nations and Metis peoples with Aboriginal and Treaty rights; (c) other licensees who may take resources, such as hunters, anglers, tourism operators, and non-timber forests product harvesters; and (d) the full gamut of non-commercial users such as photographers, hikers, campers, naturalists, and school groups who have little interest in physical consumptive uses per se but have a strong interest in maintaining access to healthy and perhaps nearby forests. Depending on the management objectives of community forests, all of these users and their rights could come to bear on publicly accessible forests. The complexity of rights issues that community forests embrace (some may say induce) figure prominently in the many chapters of this volume that present both practical—see chapters by Palmer and Smith (Chapter 2), Casimirri and Kant (Chapter 4), Harvey (Chapter 3), and Gunter and Mulkey (Chapter 8)—and theoretical—see Robinson (Chapter 12)—implications for forest resource management and development. Whether and how community forests can support the demands and pressures of all of these users and their rights, as promised, remains an important question.

The next decade of practice, research, and advocacy in community forestry will provide new experience and insights to draw upon. Using this typology to parse core values and management objectives can no doubt assist future scholars and practitioners in developing new understandings regarding the range of experiences with practice and institutional designs available to help implement community forests. As practice advances there remains a need for monitoring outcomes associated with different management objectives and approaches. We still do not possess a complete understanding of the full scale of benefits produced through community forestry. Developing further knowledge of user activities and, in particular, the benefits accrued, as well as how benefits are distributed (i.e., how much and to whom) would go a long way toward improving management but also demonstrating community forest successes. Some management entities may be "better" than others for achieving certain objectives, and so there remains an opportunity to systematically assess management outcomes from different organizational models. Community forest success involves other factors beyond that of administration and institutions, however. It would be helpful to fully understand how constellations of rights, authority, management objectives, and organizational governance actually work amid influences from endogenous and exogenous factors such as variations in institutional capacity, changes in government policy and markets, changing social attitudes regarding forests, and environmental change. The flexibility of institutional arrangements for community forestry and the implications for aspects of business, development, and local governance has yet to be fully determined (Pinkerton and Benner 2013). This framework can supplement other comparative research designs in this regard.

What this heuristic framework does not do is account for legislative, tenure, and regulatory differences that enable and inhibit community forestry. There remains much work to be done in this area. Nor does it help to sort out the complex political economy of Crown and private lands, especially in eastern Canada, where private forest land is so prevalent (i.e., the Maritimes), and what models for local involvement might emerge in the Maritime region. Intended as a descriptive classification tool, this framework is not suited to be used as an evaluation tool to judge effectiveness or efficiency in meeting sustainable forest management objectives; rather it is designed to provide assistance in cataloguing and making sense of community forests, recognizing that these forest land management initiatives present various configurations and mandates. For detailed assessment criteria specified for community forestry, see Teitelbaum (2014). There are many different kinds of community forests in Canada. Community forestry contexts and governance arrangements will continue to evolve, and so must this typology.

Notes

1 Common property must not be (but often is) confused with a fourth category, open-access property, where resources that are "unowned" are thus available for general use (i.e., no property rights exist so every user has equal right to benefit) (see Sandberg and Clancy 1996; also Luckert et al. 2011).

References

[AANDC] Aboriginal Affairs and Northern Development Canada. 2013. http://www.aadnc-aandc.gc.ca/eng/1100100025943/1100100025945. Accessed 1 June 2017.

Agrawal, A., A. Chhatre, R. Hardin. 2008. "Changing Governance of the World's Forests." *Science 320* (5882): 1460–62.

Agrawal, A., and E. Ostrom. 2001. "Collective Action, Property Rights, and Decentralization in Resource Use in India and Nepal." *Politics and Society* 29 (4): 485–514.

Allan, K., and D. Frank. 1994. "Community Forests in British Columbia – Models That Work." *Forestry Chronicle* 70 (6): 721–24.

Ambus, L., and G. Hoberg. 2011. "The Evolution of Devolution: A Critical Analysis of the Community Forest Agreement in British Columbia." *Society and Natural Resources* 24 (9): 933–50.

Beckley, T.M. 1998. "Moving Toward Consensus-Based Forest Management: A Comparison of Industrial, Co-managed, Community and Small Private Forests in Canada." *Forestry Chronicle* 74 (5): 736–44.

Beckley, T.M., and O. Korber. 1996. "Clear Cuts, Conflict, and Co-management: Experiments in Consensus Forest Management in Northwest Saskatchewan." Edmonton: Natural Resources Canada, Canadian Forest Service, Northern Forestry Centre. https://cfs.nrcan.gc.ca/publications?id=11587.

Belsky, J. 2008. "Creating Community Forests." In *Forest Community Connections: Implications for Research, Management, and Governance*, edited by E. Donoghue, and V. Sturtevant, 219–42. Washington, DC: RFF Press Book.

Benner, J., K. Lertzman, and E. Pinkerton. 2014. "Social Contracts and Community Forestry: How Can We Design Forest Policies and Tenure Arrangements to Generate Local Benefits?" *Canadian Journal of Forest Research* 44: 903–13.

Betts, M. 1997. "Community Forestry in New Brunswick." *International Journal of Ecoforestry* 12 (3): 247–54.

Booth, A.L. 1998. "Putting "Forestry" and "Community" into First Nations' Resource Management." *Forestry Chronicle* 74 (3): 347–52.

Booth, A.L., and B.R. Muir. 2013. "'How Far Do You Have to Walk to Find Peace Again?': A Case Study of First Nations' Operational Values for a Community Forest in Northeast British Columbia, Canada." *Natural Resource Forum* 37 (3): 153–66.

Booth, A.L., and N.W. Skelton. 2011. "'There's a Conflict Right There': Integrating Indigenous Community Values into Commercial Forestry in the Tl'azt'en First Nation." *Society and Natural Resources* 24 (4): 368–83.

Bullock, R. 2007. "Two Sides of the Forest." *Journal of Soil and Water Conservation* 62 (1): 12A–15A.

———. 2010. "A Critical Frame Analysis of Northern Ontario's 'Forestry Crisis.'" PhD diss., University of Waterloo.

———. 2012. "'Mill Town' Identity Crisis: Reframing the Culture of Forest Resource Dependence in Single Industry Towns." In *Social Transformation in Rural Canada: New Insights into Community, Cultures and Collective Action*, edited by J. Parkins and M. Reed, 269–90. Vancouver: University of British Columbia Press.

Bullock, R., D. Armitage, and B. Mitchell. 2012. "Shadow Networks, Social Learning, and Collaborating through Crisis: Building Resilient Forest-based Communities in Northern Ontario, Canada." In *Collaborative Resilience: Moving through Crisis to Opportunity*, edited by B. Goldstein, 309–37. Cambridge, MA: MIT Press.

Bullock, R., and K. Hanna. 2012. *Community Forestry: Local Values, Conflict and Forest Governance*. Cambridge: Cambridge University Press.

Bullock, R., K. Hanna, and D.S. Slocombe. 2009. "Learning from Community Forestry Experience: Challenges and Lessons from British Columbia." *Forestry Chronicle* 85 (2): 293–304.

Bullock, R., and M. Reed. 2015. "Towards an Integrated System of Communities and Forests in Canada." In *Community Forestry in Canada: Lessons from Policy and Practice*, edited by S. Teitelbaum. Vancouver: University of British Columbia Press.

[CCA] Canadian Co-operative Association. 2014. "Homepage." Accessed 28 May 2014. http://www.coopscanada.coop.

Charnley, S., and M.R. Poe. 2007. "Community Forestry in Theory and Practice: Where Are We Now?" *Annual Review of Anthropology* 36: 301–36.

Duinker, P., P. Matakala, F. Chege, and L. Bouthillier. 1994. "Community Forests in Canada – An Overview." *Forestry Chronicle* 70 (6): 711–20.

Duinker, P., P. Matakala, and D. Zhang. 1991. "Community Forestry and its Implications for Northern Ontario." *Forestry Chronicle* 67 (2): 131–35.

Dunkin, J. 2008. "A Forest for the Trees: Deforestation and Conservation Efforts in Northumberland County, Ontario 1870–1925." *International Journal of Regional and Local Studies* 4 (1): 47–70.

Elias, H. 2000. "Harrop and Procter: How a Persistent Community Fashioned Its Own Forestry Future." *Ecoforestry* 15: 218–26.

Furness, E., H. Harshaw, and H. Nelson. 2015. "Community Forestry in British Columbia: Policy Progression and Public Participation." *Forest Policy and Economics* 58: 85–91.

Furness, E., and H. Nelson. 2012. "Community Forest Organizations and Adaptation to Climate Change in British Columbia." *Forestry Chronicle* 88 (5): 519–24.

Government of Ontario. 1990. Conservation Authorities Act, R.S.O. 1990, c. C.27. Accessed 18 January 2017. https://www.ontario.ca/laws/statute/90c27.

Grainger, S., E. Sherry, and G. Fondahl. 2006. "The John Prince Research Forest: Evolution of a Co-management Partnership in Northern British Columbia." *Forestry Chronicle* 82 (4): 484–95.

Gunter, J., ed. 2004. *The Community Forestry Guidebook: Tools and Techniques for Communities in British Columbia*. Kamloops and Kaslo: FORREX–Forest Research Extension Partnership and British Columbia Community Forest Association.

Harvey, S., and B. Hillier. 1994. "Community Forestry in Ontario." *Forestry Chronicle* 70 (6): 725–30.

Kube, M. 2012. "Community-based Forest Management." *Forestry Chronicle* 88 (2): 101–3.

Krogman, N., and T. Beckley. 2002. "Corporate 'Bail-outs' and Local 'Buyouts': Pathways to Community Forestry?" *Society and Natural Resources* 15 (2): 109–27.

Lawler, J., and R. Bullock. 2017. "A Case for Indigenous Community Forestry." *Journal of Forestry* 115 (2): 117-125.

Leslie, E. 2016. "Stronger Rights, Novel Outcomes: Why Community Forests Need More Control over Forest Management." In *Community Forestry in Canada: Lessons from Policy and Practice*, edited by S. Teitelbaum, 311–28. Vancouver: University of British Columbia Press.

Luckert, M., D. Haley, and G. Hoberg. 2011. *Policies for Sustainably Managing Canada's Forests: Tenure, Stumpage Fees, and Forest Products*. Vancouver: University of British Columbia Press.

Markey, S., G. Halseth, and D. Manson. 2012. *Investing in Place: Economic Renewal in Northern British Columbia*. Vancouver: University of British Columbia Press.

Masse, S. 2002. "Forest Tenant Farming as Tested in Quebec: A Socio-Economic Evaluation." *Forestry Chronicle* 78 (5): 658–64.

McCullough, R. 1995. *Landscape of Community: A History of Communal Forest in New England*. Hanover, NH: University Press of New England.

Mulkey, S., and J.K. Day, eds. 2012. *The Community Forestry Guidebook II: Effective Governance and Forest Management.* FORREX Series Report No. 30. Kamloops: FORREX Forum for Research and Extension in Natural Resources and Kaslo: British Columbia Community Forest Association. http://bccfa.ca/wp-content/uploads/2013/03/FS30_web-proof.pdf.

Natural Resources Canada. 2014. "Forests in Canada." Accessed 13 February 2014. http://www.nrcan.gc.ca/forests/canada/13161.

[OMNR] Ontario Ministry of Natural Resources. 1986. *Evergreen Challenge: The Agreement Forest Story.* Toronto: Queen's Printer for Ontario.

Palmer, L., M.A.P. Smith, and C. Shahi. 2016. "Community Forestry on Crown Land in Northern Ontario: Emerging Paradigm or Emerging Anomaly? In *Community Forestry in Canada: Lessons from Policy and Practice*, edited by S. Teitelbaum, 94–135. Vancouver: University of British Columbia Press.

Passelac-Ross, M., and M.A.P. Smith. 2013. "Accommodation of Aboriginal Rights: The Need for an Aboriginal Forest Tenure." In *Aboriginal Peoples and Forest Lands in Canada*, edited by D.B. Tindall, R. Trosper, and P. Perrault, 129–48. Vancouver: University of British Columbia Press.

Pinkerton, E., and J. Benner. 2013. "Small Sawmills Persevere While the Majors Close: Evaluating Resilience and Desirable Timber Allocation in British Columbia, Canada." *Ecology and Society* 18 (2): 34.

Pinkerton, E., R. Heasliip, J.J. Silver, and K. Furman. 2008. "Finding 'Space' for Comanagement of Forests within the Neoliberal Paradigm: Rights, Strategies, and Tools for Asserting a Local Agenda." *Human Ecology* 36 (3): 343–55.

Ribot, J., A. Agrawal, and A. Larson. 2006. "Recentralizing while Decentralizing: How National Governments Reappropriate Forest Resources." *World Development* 34 (11): 1864–86.

Robinson, D., M. Robson, and R. Rollins. 2001. "Towards Increased Citizen Influence in Canadian Forest Management." *Environments* 29 (2): 21–41.

Roy, M.A. 1989. "Guided Change through Community Forestry: A Case Study in Forest Management Unit 17 – Newfoundland." *Forestry Chronicle* 65 (5): 344–47.

Sandberg, A., and P. Clancy. 1996. "Property Rights, Small Woodlot Owners and Forest Management in Nova Scotia." *Journal of Canadian Studies* 31 (1): 25–47.

Schlager, E., and E. Ostrom. 1992. "Property Rights Regimes and Natural Resources: A Conceptual Analysis." *Land Economics* 68 (3): 249–62.

Shackleton, S., B. Campbell, E. Wollenberg, and D. Edmunds. 2002. "Devolution and Community-based Natural Resource Management: Creating Space for Local People to Participate and Benefit?" *ODI Natural Perspectives* 76: 1–4.

Smith, M.A.P. 2013. "Natural Resource Co-management with Aboriginal Peoples in Canada: Coexistence or Assimilation?" In *Aboriginal Peoples and Forest Lands in Canada*, edited by D.B. Tindall, R. Trosper, and P. Perrault, 89–133. Vancouver: University of British Columbia Press.

Smith, P. 1995. *Aboriginal Participation in Forest Management: Not Just Another Stakeholder.* Ottawa: National Aboriginal Forestry Association.

Teitelbaum, S. 2014. "Criteria and Indicators for the Assessment of Community Forestry Outcomes: A Comparative Analysis from Canada." *Journal of Environmental Management* 132: 257–67.

Teitelbaum, S., T. Beckley, and S. Nadeau. 2006. "A National Portrait of Community Forestry on Public Land in Canada." *Forestry Chronicle* 82 (3): 416–28.

Teitelbaum, S., and R. Bullock. 2012. "Are Community Forestry Principles at Work in Ontario's County, Municipal, and Conservation Authority Forests?" *Forestry Chronicle* 88 (6): 697–707.

Tyler, S., L. Ambus, and D. Davis-Case. 2007. "Governance and Management of Small Forest Tenures in British Columbia." *Journal of Ecosystem Management* 8 (2): 67–79.

Vernon, C. 2007. "A Political Ecology of British Columbia's Community Forests." *Capitalism, Nature, Socialism* 18 (4): 54–74.

Waldron, J., 2012. *The Rule of Law and the Measure of Property*. Cambridge: Cambridge University Press.

Wilson, E. 1943. "Forestry in Postwar Rehabilitation." *Forestry Chronicle* 19 (1): 14–16.

Wollenberg, E., R. Iwan, G. Limberg, and M. Moeliono. 2008. "Locating Social Choice in Forest Comanagement and Local Governance: The Politics of Public Decision Making and Interests." In *Public and Private in Natural Resource Governance: A False Dichotomy?*, edited by T. Sikor, 27–43. London: Earthscan.

Westman, C.N. 2005. "Tutelage, Development and Legitimacy: A Brief Critique of Canada's Indian Reserve Forest Management Regime." *Canadian Journal of Native Studies* 25 (1): 207–14.

Wyatt, S. 2008. "First Nations, Forest Lands, and 'Aboriginal Forestry' in Canada: From Exclusion to Comanagement and Beyond." *Journal of Canadian Forest Research* 38: 171–80.

Transformative Community Organizing for Community Forests: The Northern Ontario Sustainable Communities Partnership

Lynn Palmer and M.A. (Peggy) Smith

The Northern Ontario Sustainable Communities Partnership (NOSCP) was established in 2006 as an inclusive, grassroots group in response to the "forestry crisis" that was rippling across the major forest product–producing provinces in Canada. NOSCP viewed the crisis in northern Ontario as an opportunity to re-evaluate the structure of the forest sector and to focus on solutions to achieve long-term sustainability for local communities and the Crown forests upon which they depend. NOSCP aims to promote a regional approach to the implementation of community forests. The group advocates for forest tenure policy reform to enable implementation of community forestry and supports communities in their efforts to develop community forest initiatives.

This chapter describes the transformative approach to community organizing undertaken by NOSCP to promote amelioration of the historic command-and-control forest tenure system to support community forestry in northern Ontario. This system alienated municipalities and Indigenous communities from forest management decision making, leading to dependent, unstable local economies, thus compromising community well-being. NOSCP uses transformative community organizing (TCO) to create a collective voice that builds power among northern Ontario citizens to represent their own interests in forest management and to raise critical consciousness about the need for an alternative forest tenure approach that supports community forests. This approach challenges the assumptions of the dominant forestry system that

NOSCP sees as part of the capitalist model of neoliberalism. In concert with the TCO approach, power, as it relates to forest tenure within the social and historical context in northern Ontario, is the key issue addressed by NOSCP.

This chapter takes the perspective that transformative community organizing by NOSCP is a social change movement. This movement is part of a wider force for social and political justice that has emerged to contest the neoliberal political-economic paradigm that has dominated industrialized nations for the past three decades. A resurgence of grassroots social movements, not seen since the pre-neoliberal era, began with an outbreak of the alter-globalization movement in the 1990s[1] (Chesters and Welsh 2005; Engler 2007) and has since expanded to worldwide protests by the Occupy movement and others over climate change and social justice (Harden-Donahue 2014; Rehmann 2012; Sharlet 2013). In Canada, social movements critical of resource extraction, such as tar sands expansion and fracking, have gained prominence (Schwartz and Gollom 2013; Loney 2013; Crawford 2014; Ruiz Leotaud 2014). These Canadian movements include Idle No More, which arose in 2012 to protest federal weakening of environmental regulations and recognition of Indigenous rights given the unequal position of Indigenous peoples throughout the country (Jarvis 2013). The Idle No More movement is spurred on by legal wins that acknowledge constitutionally recognized Aboriginal and treaty rights related to lands and natural resources (Gallagher 2012; Hildebrandt 2014).

These contemporary social change movements have used the global economic crisis that has shaken neoliberal capitalism as a window of opportunity to promote their social change agendas. NOSCP similarly capitalized on the forest sector downturn as an opening to push for transformation of the forestry system in northern Ontario to one that supports community forestry. This direction aligns with worldwide community forestry social movements that have been instrumental in achieving local and Indigenous rights over forest resources in recent decades; their calls for greater tenure and usufruct (property use) rights continue to intensify (Taylor et al. 2013).

In addition to being a social change movement, NOSCP is also described as an informally structured community organization that promotes community forests as an alternative development model to the dominant neoliberal forestry approach. NOSCP has successfully united these two traditions of community organizing.

The chapter begins with an overview of concepts and theories on neoliberalism and TCO to provide an understanding of (1) the broader political-economic paradigm that has influenced the forestry system in northern

Ontario, and (2) the approach NOSCP has used to promote transformation of this system. The northern Ontario forestry system is next discussed to illustrate the context within which NOSCP operates. The chapter then traces the evolution of NOSCP from its inception to the present, with a discussion of the action and education activities it has undertaken during that time as both a social movement and a community organization.

Neoliberalism: Impacts and Alternatives

While there are various understandings of neoliberalism, many which are informed by Marxism or the theories of Foucault,[2] the different perspectives all share a common concern about power relations being the central problem associated with capitalism (Springer 2012). Neoliberalism is founded in classical liberalism, which sees government control of economic decision making through central planning as a loss of freedom. The key focus of neoliberalism is the centrality of the market. Neoliberalism embraces neoclassical economics, the dominant school of economic thought that links supply and demand to the rational choices of self-interested individuals and their ability to maximize utility or profit based on tastes and preferences (Harvey 2010).

Neoliberal ideology purports to embrace "free" markets—free from state interference—as the optimal mechanism for economic development (Brenner and Theodore 2002). Various accounts convey how neoclassical economic theory continues to dominate mainstream economics despite extensive evidence that has shown its failings and the admissions of mistaken beliefs by some former staunch supporters following the global economic crisis (Beinhocker 2006; Clark and Treanor 2008; Posner 2009; Cassidy 2010; van der Veen 2013; Mirowski 2014).[3] In their broad social and economic critique of neoliberal capitalism, DeFilippis et al. (2010) describe how the larger political-economic processes associated with this approach have resulted in a "global economic tsunami" fraught with frequent bust cycles that have had extreme repercussions for the poor and politically marginalized. These authors discuss the negative impacts of market expansion into the social sphere stemming from decentralization of the state, where a greater burden has been placed on communities without increasing their authority by moving key economic decisions further away from local control.

While DeFilippis et al. (2010) argue for some degree of decentralization to enhance local democracy, they stress that it should not replace the role of the state as the locus in society that has the power to redistribute wealth and limit the power of capital. Their approach supports *democratic* decentralization, which is defined as arising due to a demand for participation from below through social movements and local governments that challenge the traditional, centralized approach to public policy (Conyers 1983; Agrawal and Ostrom 2001; Larson 2005; Larson and Soto 2008). Democratic decentralization ideally results in the formation of autonomous, local governments and discourse about participation in decision making, participatory democracy, pluralism, and rights (Conyers 1983).

Alternatives to Neoliberal Capitalism

A popular call from the inaugural World Social Forum in 2001 (WSF 2015) advocates for building economic and social spaces beyond the dominance of neoliberal capitalism that would see a shift to smaller enterprises rooted in communities and more collective ownership (Cavanagh and Broad 2012; van der Veen 2013). These alternatives are intended to better serve people and the planet instead of generating profits for the few. Leyshon and Lee (2003) argue that openings for alternative economic spaces exist where the network of capitalism is weak. Such openings provide potential for a diversity of economic spaces that function from a different perspective than capitalism. Participatory experiments that promote a broader understanding of economic practice, based on different values and approaches to exchange, have already managed to carve out such spaces within the neoliberal context (Gibson-Graham 1996, 2005, 2006, 2008; van der Veen 2013). These include fair trade commodities (Mutersbaugh et al. 2005; Taylor 2005) and co-operatives (Gibson-Graham 2006; van der Veen 2013; Novkovik and Webb 2014). McCarthy (2006a) asserts that the recognition of and search for more co-operative forms of economic and social organization is a vital political act.

In the realm of forestry, community forests that emerged due to popular demand in the 1990s on Crown lands in British Columbia have recently been analyzed as political and economic alternatives to the dominant neoliberal approach of centralized state control and the industrial forestry model. McCarthy (2006b) suggests that despite the small scale of BC's community forests, they nevertheless might be a wedge for a more democratic and sustainable future. Pinkerton et al. (2008) illustrate what they consider to be a successful example of a BC community forest—the Harrop-Procter Community Forest that operates as a co-operative—having created space within a neoliberal policy context to assert community values, goals, and strategies to attain a real voice in forest management. More recent accounts of this same community forest are given in this volume (by Egunyu and Reed, Chapter 9; Leslie, Chapter 10). Robinson

(Chapter 12) describes how community forestry, which is generally thought to be a more equitable and environmentally sound approach than large-scale industrial forestry, is also an economically superior model. Robinson therefore asserts that community forestry is superior to the conventional approach in all pillars of sustainability.

Transformative Community Organizing

Rubin and Rubin (1992) outline several important elements of community organizing, including: social power gained through collective action; learning how power operates; capacity for democracy; and sustained social change as an outcome. Transformative community organizing (TCO) mobilizes citizens through consciousness raising to demand fundamental social change in order to transform the dominant system. This approach aligns well with complexity theory, which focuses on the transformation of complex adaptive systems that have the capacity to adapt in a changing environment into new, more resilient configurations (Gunderson and Holling 2002; Walker et al. 2004; Walker and Salt 2006).[4]

The core focus of TCO, also called radical (Reisch 2013) or opposition/action (Shragge 2013) organizing, is power—the identification of who has it and how it is used to maintain the system within its existing economic and political context. TCO focuses on transferring power from government and the market to community. A second aspect of TCO concerns values and ideology: these focus on increasing social and economic equality and extending democracy based on the underlying principles of justice, equity, respect, and diversity. TCO seeks to achieve fundamental and sustained social change that results in a more democratic and participatory system based on economic and social justice. This occurs by first acknowledging and then challenging power, followed by a major redistribution of power and resources to create alternative institutions based on democracy and direct control by citizens. Widespread consciousness raising through TCO provides citizens with an understanding of how the dominant system works, who holds power, and why it is necessary to build power to create social change. A network of citizens is then mobilized to undertake collective action to challenge the legitimacy of the dominant power relations and the interests they serve to create an alternative political and socio-economic culture. The political-economic perspective taken by TCO involves a critical analysis of the root causes of social problems as they relate to the fundamental distribution of resources and power in dominant systems and the development of strategies for their transformation (Reisch 2013;

DeFilippis et al. 2010; Shragge 2013). This analysis recognizes that history, culture, and context are significant factors in the creation of social problems and are equally important for solutions.

The practice of TCO utilizes a range of strategies and tactics to build an understanding among individuals and communities about the existing context and its limitations—the workings and power relations of the dominant system—and the necessity to challenge these through collective action to build long-term, positive social change. Action and education are two key strategies for the practice. DeFilippis et al. (2010) emphasize how popular education, or "education for critical consciousness" (Freire 1974), is an important aspect of organizing to enable understanding of contemporary processes of neoliberalism and capitalist globalization. The work of organizing can include opposition to policies that allow oppression and inequality, as well as support for local, often smaller-scale alternative institutions that exhibit new kinds of economic and social relations more equitable than those of the dominant paradigm. Because of the political goals and analysis of social and economic inequality, resistance and conflict are emphasized at the core of TCO activities; however, this conflict orientation does not imply constant conflict but rather the recognition of its potential even if it is rarely necessary, along with a willingness to engage in explicitly political practices (DeFilippis et al. 2010; Shragge 2013). In this way, TCO comes from a position of power and opposition to anti-democratic forces rather than acquiescence to so-called "partners" that in reality hold the bulk of power. This notion of TCO coincides with democratic decentralization in forest management that sees genuine representation based on accountable local authorities able to make and implement decisions (Ribot 2002; Ribot et al. 2006).

Most transformative social change has been the product of social movements that have organized and mobilized local communities to challenge oppression and injustice and expand political, social, and economic democracy (DeFilippis et al. 2010). A social movement is defined as a network of activists and organizations that are loosely affiliated around a common purpose to undertake collective action (Della Porta and Diani 2006; Diani 2011; Staggenborg 2011). An emerging scholarship on social movements that uses complexity theory as a lens conceptualizes social networks as complex assemblages of actors, discourses, alliances, interests, and knowledge (De Landa 2006). Social movement networks viewed through this lens, such as those associated with the alter-globalization movement, are considered to be emergent, diverse, and self-organized through democratic, bottom-up processes (Chesters and Welsh 2005; Escobar 2008; MacFarlane 2009; Rankin and Delaney 2011).

DeFilippis et al. (2010) and Shragge (2013) make a link between social movements and community organizations despite their different histories and orientations, which have led to them typically being viewed as distinct traditions with different objectives. Social movements are generally informally organized efforts that are without formal structures and inherently unstable and episodic, with beginnings, peaks, and declines. Community organizations, in contrast, tend to focus on building an organizational structure to deliver needed services or complete projects.

Northern Ontario Context: People, Place, Forestry System

Northern Ontario is a vast region that occupies 80 percent of the province's land mass but has only 8 percent of its population (approximately 800,000 people). Most of the population is concentrated near the southern boundary, which runs approximately from the Mattawa River in the east, across Lake Nipissing, and to the French River in the west, as well as in five major urban centres: North Bay, Sudbury, Sault Ste. Marie, Thunder Bay, and Timmins. Communities in the region include both municipalities under the jurisdiction of the province and First Nation reserves under the jurisdiction of the federal government. Municipalities, which include the Metis population for statistical purposes, are commonly single-industry towns with few employment options other than in the resource sectors, including forestry. First Nation communities have historical and contemporary ties to their traditional territories, which encompass large areas of Crown forest lands owned by the province and licensed by it for resource extraction (Smith 1998). These provincial Crown lands are subject to historic treaties and place a burden on the provincial Crown to protect treaty rights, such as hunting, fishing, and trapping, in the face of any resource development (Gallagher 2012).

The Crown forests of the region are located in the boreal and Great Lakes–St. Lawrence forest zones. A significant portion of this forest area was defined as the "Area of the Undertaking" (AOU) for the Class Environmental Assessment for Timber Management on Crown Lands in Ontario conducted from the late 1980s to 1994 (EA Board 1994). The AOU's northern boundary approximates the 50th parallel, with its southern boundary running from the Mattawa River in the east, across Lake Nipissing, and to the French River in the west. The AOU is the geographic focus of this chapter.

Forest management on Crown lands in northern Ontario has historically functioned as a centralized, command-and-control system through the Ontario Ministry of Natural Resources and Forestry (OMNRF).[5] The province licenses timber from Crown forests to commodity forest industries (pulp and paper, dimensional lumber) that focus on export, primarily to the United States. With a minimal diversity of actors and forest products, the forest management system has emphasized economic production and scientific management to supply timber to the forest industry (Burton et al. 2003) consistent with neoclassical economics embraced by neoliberalism.

This centralized command-and-control system, which fits the "staples" model of economic development based on resource extraction and resource commodity export, has provided little room for local decision making (Thorpe and Sandberg 2008). Although a community forest program created by a short-lived progressive provincial government in the early 1990s saw the implementation of four community forest pilots,[6] by the late 1990s, the intensification of neoliberalism throughout Canada had a major influence on forest policy in Ontario's Crown forests. As a result, the province increased its focus on industrial tenures (OMNR 1998) and the dominant forestry regime prevailed, subjecting both the forest industry and the communities that depended on it—primarily the municipalities—to the boom-and-bust cycles associated with staples commodity markets (Clapp 1998).[7]

The inherently unstable approach of Ontario's Crown forestry system eventually led to a major downturn experienced by the forest industry in the new millennium. The downturn worsened and culminated in a forestry crisis that saw an unprecedented number of mill closures from 2005 to 2006 (OFC n.d.) with significant negative socio-economic impacts in municipalities throughout the region (Bogdanski 2008; Patriquin et al. 2009). While First Nation communities also experienced some negative impacts due to the crisis, having been largely excluded from the forest-based economy from the outset and the benefits it extracted from their traditional territories, they have always faced much greater economic challenges than municipalities in the same region (Southcott 2006).[8]

With the advent of the forestry crisis, the province first attempted a forest licence conversion program in 2007 from single entity to "co-operative" licences aimed to increase efficiency. However, by the fall of 2009 Ontario recognized that the forestry crisis could not be resolved by this modest adjustment to the existing system. Amid mounting pressure for measures to address the forestry crisis from local communities and First Nations, as well as the forest industry, OMNR initiated an unprecedented forest-tenure reform process in September 2009. The process continued until May 2011 and involved a number of public and Indigenous consultations throughout northern Ontario. A widespread call came from communities for community forestry early in the tenure reform process (Speers 2010), and numerous communities developed community forest proposals for implementation under a new forest tenure system.

A significant outcome of the reform was the Ontario Forest Tenure Modernization Act, 2011. The Act outlines how one new tenure model, the Local Forest Management Corporation (LFMC), a Crown corporation, will function. The Act also enabled the establishment of two of these models. New policy was also created at that time for a second model, the Enhanced Sustainable Forest Licence (ESFL) (OMNDMF 2011). The province stated that these new tenure models were designed to increase opportunities for local and Indigenous community involvement and forest-sector competitiveness (OMNDMF 2011), although it was not made clear if and how they would accommodate Aboriginal and treaty rights.

The first LFMC was established in April 2013, and a number of ESFLs are currently under development. Various groups of communities, including both First Nations and municipalities, continue to propose and develop collaborative, regional community forestry initiatives that they aim to see implemented under the ESFL tenure option. A review of all forest tenure models by the province commenced in 2016.

Since tenure reform is ongoing, it is unclear at this point if the forestry system will undergo a transformation that would see forest tenure policy that supports community forests or if the status quo will reassert itself as a manifestation of the broader neoliberal regime. Palmer et al. (2016) evaluate northern Ontario's forestry system through the lens of complexity theory, where the forestry crisis is characterized as a system collapse. Tenure reform is characterized as a phase of subsequent system reorganization that offers the potential for system transformation to a more resilient configuration consistent through the concept of transformative community organizing.

Evolution of NOSCP

At the height of the forestry crisis, the concept for NOSCP arose out of discussions among a group of participants from academia, NGOs, First Nations, and municipalities at the September 2006 Lakehead University Biotechnology Symposium. These individuals organized or participated in several symposium sessions about the need for diversification of forest products and new forest tenure policy to foster the transition to a sustainable bio-economy in northern Ontario given the forestry crisis. The founding meeting of NOSCP was convened shortly afterward in Thunder Bay with these and additional participants who came together to discuss concerns about the forestry crisis and the need for forest tenure that provides northern Ontario's residents with greater rights and responsibilities over public forests in order to achieve community sustainability.

Subsequent meetings of a similar ad hoc, inclusive nature ensued on a regular basis over the next year. Participation expanded to include representatives from several academic institutions, First Nation communities and organizations, municipalities, provincial and federal governments, NGOs, unions, and individual citizens from throughout northern Ontario. These meetings rotated among different host organizations located in Thunder Bay, and all involved teleconferencing to allow participation from anywhere in the province. Capacity for meeting coordination and minutes was provided for the first year and a half by a staff person at the Thunder Bay office of the Canadian Parks and Wilderness Society–Wildlands League.

The range of individuals, community representatives, and organizations that became involved early on in NOSCP reflects the stance of DeFilippis et al. (2010) and Shragge (2013), who contend that to challenge the power structures of contemporary capitalism, linkages beyond the local are essential through broad alliances to address problems caused by forces and decisions that transcend any individual community. NOSCP was accordingly established as a regional social change movement focused on building alliances among communities and organizations throughout northern Ontario.

In addition to emerging as a social movement to promote community forestry, NOSCP was established as a transformative community organization with a name, mission statement, principles, and goals that were determined soon after its inception. NOSCP united these two traditions for a common cause (DeFilippis et al. 2010) in order to recognize their common origins and elements and foster having them perceived as parts of the same broader struggle.

NOSCP's mission clearly states the need for social change to transform the dominant economic system to one that is just and sustainable for both the communities and forest ecosystems in northern Ontario. NOSCP principles reflect the fact that the needed transformation requires a holistic view of the land rather than a focus on any single resource and that NOSCP will always focus on long-term, proactive solutions to achieve sustained social change rather than those that are short-term or reactionary. NOSCP established two different but related goals. The first is to promote sustainable community bio-economic development through diversification of northern Ontario's forest economy based on value-added production of both timber and non-timber products. The second goal focuses on the need to transform the forest tenure system to achieve a more democratic and participatory forest management system that provides greater benefit to northern Ontarians. The mission, principles, and goals of NOSCP clearly articulate a critical analysis of the root problem(s), an explicit commitment to promoting social and economic justice, and an alternative direction wherein hierarchy and domination are ended. All of these are regarded as essential to building a wider oppositional culture to the existing power relations of the dominant paradigm through TCO (DeFilippis et al. 2010; Shragge 2013).

NOSCP was created with an informal and decentralized organizational structure that minimizes bureaucracy and formal leadership. Northeastern and northwestern Ontario co-chairs are based at Lakehead and Laurentian Universities so as to have a presence in each region through neutral organizations not directly involved with forest tenure. There is no membership fee, and any individual or organization can be a member as long as they agree with NOSCP's mission, principles, and goals. This approach has created a democratic space based on direct participation in the organization, where those who show up make the decisions. Shragge (2013) notes the advantages of maintaining an informal approach to community organizing—it requires few resources and maintenance of the organization is easier—and the disadvantages associated with formalization—institutionalization, professionalization, depoliticization, and demobilization—that tend to produce a shift away from mobilizing citizens for a greater service orientation.

Early Organizing

A number of action and educational activities were undertaken by NOSCP during its first year to generate awareness about the group and to begin to work toward achieving its goals. In addition to regular tele-meetings, presentations were given to various community groups as well as OMNR. Several videoconferencing sessions were offered via the Northern Ontario Medical School facilities on topics related to non-timber forest products to promote a diversified northern bio-economy. An additional public videoconference was offered at Lakehead University in June 2007 to initiate dialogue about the need for forest tenure reform, with presentations from OMNR, a local First Nations organization, and an economist. "Community-based Forest Management for Northern Ontario: A Discussion and Background Paper" (NOSCP 2007a) was jointly written by several participants and presented at the videoconference.

With the ever-prevalent negative socio-economic impacts of the forestry crisis undiminished throughout the region of northern Ontario by the second half of NOSCP's startup year, participants agreed to concentrate their energies on the second goal of promoting forest tenure reform that would support community forests, concurring that tenure was the greatest barrier and opportunity with respect to achieving both goals. Efforts thus shifted to developing the Northern Ontario Community Forest Charter to promote community-based decision making for the publicly owned forests of northern Ontario (NOSCP 2007b). The charter, which became the guiding document for NOSCP's subsequent activities, was released at the group's first press conference in the fall of 2007 (Brown 2007) and was subsequently distributed for endorsement.

The charter's twelve principles broadly address good governance, shared decision making, separation of forest management from any one specific user group (i.e., a forestry company), the promotion of a diverse forest sector through best end use of forest resources with a focus on value-added production of both timber and non-timber values, less reliance on commodity industries, ensuring that local communities benefit from forest development, upholding Aboriginal and treaty rights, and fair trade. The charter adds depth to NOSCP's critical analysis of the root problem related to the forest system, the need to redistribute forest resources to benefit northern Ontario communities and to address power relations to do so. It also makes explicit the vision for an alternative in the form of community forests guided by the principles. Endorsement of the charter by a wide range of individuals and organizations began soon after its release.

A key aspect of the charter is that it commits not only to respect but also to "help resolve" the outstanding issues around implementation of Aboriginal and treaty rights in forest management. This charter principle was developed with the understanding that this is a crucial aspect of a truly democratic forest management system in Ontario. NOSCP's position on this issue is in alignment with both the improving legal climate in Canada regarding Aboriginal and treaty rights (Gallagher 2012) and a new wave of social change that is working to address the unequal position of Indigenous peoples, as seen in the Idle No More movement.

Also created during NOSCP's startup year was a website hosted by the Geraldton Community Forest, a contracting company in the Municipality of Greenstone in northwestern Ontario that originated as one of the community forest pilot projects of the 1990s. The website enabled outreach and generated awareness about the establishment of NOSCP, consciousness raising about the organization's focus to transform the dominant forest tenure system, sharing of information about NOSCP activities and publications, and broad distribution and endorsement of the charter.

A two-day workshop was held in Thunder Bay in March 2009 to provide an inaugural forum in northern Ontario for information sharing and participatory learning among a wide range of participants about how local communities might gain more control over decision making about local forests. In order to obtain workshop funding, NOSCP obtained a business licence while maintaining its informal structure. NOSCP gained support from OMNR as a workshop partner, both as a presenter and funder. This strategy employed the principle that the state should be a facilitator of progressive change for the common good rather than a supporter of the status quo (Ferge 2000).

The workshop was a resounding success, with participation from numerous First Nations as well as local municipalities, local citizens committees, academics, government, environmental non-governmental organizations, the forest industry, and unions (NOSCP 2009). The concept that "we are all treaty people" who need to revive the spirit and intent of the treaties to share lands and resources was an underlying workshop theme. Although First Nations and other communities had previously raised concerns about the forest tenure system and the need for greater input by communities, this forum initiated widespread mobilization of northern Ontario residents toward change in the dominant forestry system; pressure was put on OMNR in this public setting for reform of the tenure system to support community forests.

Critiquing Tenure Reform

Various members of NOSCP participated in the public and Indigenous tenure reform consultations that were held from the fall of 2009 to the spring of 2010 wherein a consistent message was promoted about the need for a new tenure system that supports community forests. Shortly after the release of a draft proposal for a new provincial tenure system (OMNDMF 2010), NOSCP developed and submitted to OMNR a commentary with recommendations for an alternative tenure framework based on the charter principles (NOSCP 2010). The commentary called for meaningful input to be accepted from communities about the design of localized and diverse tenure models that would ensure effective local and Indigenous community representation through democratic decentralization (Ribot 2002).

Despite NOSCP's efforts, strong lobbying by the Ontario Forest Industries Association, which represents a large segment of the forest industry, led to decisions that the new forest tenure framework would maintain existing wood supply commitments and ensure a "measured and moderate" approach to any change in the tenure system (OFIA, NOACC, NOMA, and FONOM 2011). Upon the creation of the Forest Tenure and Modernization Act in May 2011, NOSCP issued a press release (NOSCP 2011) that characterized the new legislation as a "timid beginning" that provides some tools for communities to move closer to a community forest model. Concerns raised about the changes were: (1) no provisions for community forests in which local northern Ontario communities have decision-making authority over the use and future of local forests in keeping with democratic decentralization; (2) the failures to acknowledge and provide for Aboriginal and treaty rights; (3) the province's significant control over LFMCs since they are Crown corporations; and (4) the continued sole focus on industrial timber production with no attention given to broader values based on non-timber forest products.

Given that the OMNR established Forest Industry and Aboriginal Working Groups to participate in the development and implementation of the new forest tenure system, NOSCP advocated for a parallel group to represent communities. As a result of this effort, OMNR agreed in 2011 to establish a Community Working Group with NOSCP representation to develop implementation guidelines for Enhanced Sustainable Forest Licences. As of 2013, OMNR discontinued the three working groups and created an Oversight Group that included representation from Indigenous, community, and industry sectors; however, NOSCP was not represented.

NOSCP's challenges to government and industrial perspectives on tenure reform reflect an approach to TCO that moves beyond the bounds of small-scale reform to transformational change that redefines systemic problems and challenges power relations. NOSCP characterized the new LFMCs as achieving only small-scale reform and not the needed transformation of the system that would enable community forests. This perspective supports the stance (Shragge 2013; DeFilippis et al. 2010) that state-shaped organizations can achieve some positive gains without challenging the relations of domination and power that keep the system working in the interest of neoliberal capitalism; such organizations are system-maintaining and do not achieve system transformation. Based on the guiding ESFL principles developed with input by NOSCP, it appears that this new tenure model may have greater flexibility in forest governance compared to LFMCs. Since all ESFLs are still under development, with none yet implemented, this outcome remains to be seen. Unlike LFMCs, ESFLs can include forest industry representatives in their governance structures. Partnerships between industry and communities, as are being negotiated for several developing ESFLs (see Lachance, Chapter 5, for an example), are intended by community participants to enable strong community representation in new tenure models. Yet, the very presence of forest industry at any level in the new governance structures will preclude democratic decentralization (Ribot 2002), which can occur only within the election (not government appointment) of community representatives.

Broadening the Movement

Following the creation of the new Ontario forest tenure framework, NOSCP was involved in three further events to advance community forestry as a model for collaborative decision making and development: (1) a by-invitation workshop at Lakehead University in May 2011 (Palmer et al. 2012); (2) a national, interdisciplinary conference held in Sault Ste. Marie, Ontario in January 2013 (Palmer et al. 2013); and (3) a symposium at the University of Winnipeg in June 2014 (Bullock and Lawler 2014). For all three events, NOSCP partnered with academic institutions: Lakehead University for the workshop, Algoma University and its Northern Ontario Research, Development, Ideas and Knowledge Institute for the conference, and the University of Winnipeg and its Centre for Forest Interdisciplinary Research for the symposium. Such partnerships enabled access to funding from the Social Sciences and Humanities Research Council of Canada. These events served to broaden alliances, regarded as crucial in order to contest power in the dominant system (Shragge 2013; DeFilippis et al. 2010).

A key goal of these events was to build alliances by bringing people together to share information and experiences and to develop solutions. Each involved providing funding support for participation by community representatives, presentations by community forestry researchers and practitioners from various jurisdictions, group discussions, student posters, and live streaming.

A new national network, Community Forests Canada (CFC), was established as an outcome of the Sault Ste. Marie conference to support existing and proposed community forest initiatives, policy engagement, and research throughout the country. NOSCP committed to participate in this network as a means to further an alliance to support the advancement of community forestry in Ontario and across Canada. An additional outcome of the conference was the concept for this volume. As a partner in the University of Winnipeg symposium, NOSCP helped to further outreach about the concept of community forestry in Manitoba (which has had no historical involvement with this approach) through Manitoba representatives' participation on a national panel.

These public outreach events provided the opportunity to deepen the critical analysis of the issues and potential solutions regarding forest tenure by bringing various actors together, a key aspect of TCO. Alliance building regionally, among provinces, and internationally as a result of these forums also demonstrates a key aspect of TCO: to be a force for social change, local mobilization must take place in conjunction with similar organizations and movements elsewhere to build solidarity for a wider oppositional political culture (DeFilippis et al. 2010; Shragge 2013). Alliances like Community Forests Canada can build federated structures that develop "associated democracy." The ultimate aim of this national network is to strengthen the push for improved legislation and policy to support community forestry across Canada. The alliances resulting from NOSCP's activities thus far have helped Ontario to recognize that interest in and support for community forestry is not a localized anomaly but widespread in northern Ontario and beyond.

NOSCP as a Transformative Community Organization and Social Movement

NOSCP advocates for community forests in northern Ontario through transformative community organizing based on the view that sustainable forests and democratic control of forest management lead to local benefits that create resilient forest-based communities. Its organizing approach focuses on a deep analysis of the root cause of the forestry crisis that recently affected the region. This has involved a strong critique of the command-and-control forestry system that has negatively affected forest-based communities in a region that historically has lacked a voice in that system. Also addressed by the analysis is the longer-standing issue of Indigenous marginalization due to colonization.

NOSCP's demands for social change in the forest tenure system in the form of community forests based on participatory democracy focus on securing meaningful decision making by local and Indigenous communities. NOSCP has an activist character with a radical stance and conflict perspective evident in its explicit critique of the mainstream dominant forestry paradigm affiliated with neoliberalism and the offer of an oppositional alternative. Conflict is central to its practice, which is grounded in an analysis of the political economic system in northern Ontario. This analysis guides NOSCP's commitment to progressive social change through popular mobilization of citizens in northern Ontario. NOSCP recognizes that its approach to community organizing is a means to address the basic inequalities of power that have been entrenched in the forest-based communities and First Nations in this region as a result of the historical forest tenure system.

A key aspect of NOSCP's advocacy approach has been to unite like-minded individuals and organizations throughout northern Ontario in a common political cause and vision to achieve transformation of the forest system through strong collective action to support community forests. Because forest tenure policy is under provincial jurisdiction and thus affects the entire region of northern Ontario, it is important that power to promote an alternative community forest system is built at this scale. At the same time, NOSCP works to build solidarity through alliances beyond this region that extend throughout Canada and internationally. This approach is consistent with TCO practice that works beyond the local level to achieve a wider oppositional political culture.

NOSCP uses conflict strategies and tactics to promote an understanding that the dominant forestry system, and its associated power relations, is the major cause of the current forest sector crisis in northern Ontario. Although NOSCP has not utilized direct action as a strategy—protests, demonstrations, sit-ins, occupations, or blockades that aim to directly stop or encourage specific action by their targets (Smith 2014)—its conflict orientation is nevertheless evident in its willingness to engage in explicitly political practices.

Given its structure and how it functions, NOSCP is both a community organization and social movement, with elements of both these traditions, which have often been regarded as distinct. While NOSCP has maintained an informal structure similar to that of most social movements, its existence for the last decade, together with its long-term vision for social change, displays the characteristic stability of a community organization rather than the typically episodic nature of a social movement.

NOSCP is a community organization based on the active participation of its members. This membership is informal, made up of individuals and organizations affected by the same issue who have come together to find solutions to change the system. The mission, principles, goals, and charter convey fundamental opposition to the power relations inherent in the conventional forest tenure system and reflect a long-term orientation toward an alternative future where communities are empowered and resilient through meaningful decision making about local forest management. The combination of action and education activities undertaken by NOSCP illustrates its practice of transformative community organizing. All activities have capitalized upon the period of the forestry crisis and subsequent window of tenure reform to push for transformative change of the historical forest tenure system.

Through its advocacy work, NOSCP became a force for social change. This impact is reflected in the consciousness raising that occurred among citizens in northern Ontario about the need for fundamental restructuring of the forestry system. This awareness extends to the provincial government, which has oversight for forest tenure policy, as seen in the support for the Ontario public outreach events and including NOSCP as a member of the Community Working Group that developed ESFL implementation guidelines.

Conclusions

NOSCP has made a number of significant contributions during the last decade as a transformative community organization and social movement. Through a variety of action and education activities throughout this period, NOSCP has been effective in: (1) raising consciousness and providing popular education about community forests as an alternative approach to the industrial forest model that can better serve communities and foster their resilience; and (2) influencing forest tenure policy in Ontario to move closer toward this alternative model.

The context has changed to some extent in northern Ontario since the establishment of NOSCP. The forest sector has recovered in part and is expected to rebound further, at least in the short term. Tenure reform is underway and at the implementation stage following several years of public and Indigenous consultations and the creation of new tenure legislation and policy. Yet the issues NOSCP is most concerned with still remain—how northern Ontario communities can become resilient for the long term and how Aboriginal and treaty rights can be addressed in relation to the forestry system. Although local and Indigenous communities have achieved varying degrees of input into the development of new forest tenure models through the new forest tenure policy framework, and while there appears to be greater flexibility with ESFLs compared to LFMCs, the current tenure options are limited to just these two approaches. Other limitations relating to the new tenure framework include its continued focus on industrial timber production rather than a broader range of values and the continued substantial influence of some forest industries over the tenure reform process. Communities continue to raise concerns about the ability of these models to serve their interests. Some First Nations in particular continue to propose new models tailored to their unique local circumstances. It is also unclear at this point how well the new provincial Oversight Group assisting with tenure implementation and evaluation will support community aspirations.

The significant changes in the broader political economy both within Canada and globally since the inception of NOSCP provide an updated context that will influence its future role and direction as a community organization and/or social movement. The neoliberal regime that has dominated Canada and much of the world for the past three decades persists, and free market ideology still predominates. At the same time, while the widespread reign of neoliberalism weakened social movements during the early stage of this period, these movements have since gained momentum as citizens respond in increasingly sophisticated ways to reclaim space defined by neoliberal discourse (Smith 2014). Contemporaneous with these expanding social movements are increasingly favourable Supreme Court of Canada decisions on Aboriginal and treaty rights, most significantly the Tsilhqot'in decision on Aboriginal title in 2014;[9] these decisions greatly increase the promise of Indigenous peoples regaining their place as rightful stewards of their traditional lands and obtaining meaningful decision making in natural resource development (Anderson 2014). This legal trend signals favourable prospects for the development of community forestry, whether solely by Indigenous communities or as cross-cultural collaborative efforts with non-Indigenous communities.

Community organizing is shaped by changes in the broader political-economic context within which it is embedded. These changes create opportunities as well as constraints. Consequently, as DeFilippis et al. (2010) point out, while we have not yet seen neoliberalism replaced, or a non-capitalist political economy emerge from the global economic crisis, there nevertheless remains extraordinary potential for positive social change. The key, however, is the ability for people to engage in collective action while the window of opportunity remains open. Ontario remains hesitant about devolving full control of forest management decision making to local communities. Tenure reform has not yet resulted in community forests as envisioned by NOSCP. However, the recent political, economic, and regional contexts have created a greater climate of potential to help NOSCP be a driver of social change while continuing to exert influence to improve the outcomes of forest tenure reform. With continuing economic uncertainty, pressure to uphold Aboriginal and treaty rights in natural resource development, and ongoing public concern about environmental protection, NOSCP has the opportunity to continue its role in further pushing the forest system toward a regime shift. This transformation would see the creation of forest tenure policy that is supportive of community forests as a genuine alternative to neoliberal capitalism, one which fosters a resilience approach for local communities and the forest ecosystems upon which they depend.

Whether NOSCP continues to be a social movement or shifts to a community organization that provides support for future community forests depends largely on the final outcome of forest tenure reform in Ontario. In this regard, NOSCP is still a work in progress. Its future direction will be determined based on how well it can meet its goals in a changing political-economic context.

Notes

1 The rise of the alter-globalization movement is chronicled by Chesters and Welsh (2005), beginning with the Intercontinental Gatherings for Humanity and Against Neoliberalism (Zapatista Encuentros) in Mexico in the mid-1990s, followed by protests against global financial institutions (International Monetary Fund, World Trade Organization) later that decade, and the establishment of the World Social Forum in 2001 and subsequent regional sub-conferences by social movements opposed to neoliberalism and the domination of the world by capital and any form of imperialism.

2 Michel Foucault was a French philosopher, historian of ideas, social theorist, and literary critic. His theories addressed the relationship between power and knowledge, and how they are used as a form of social control through societal institutions. In one of his well-known critiques, Foucault (1982) describes how neoliberal *subjectivation* affects individuals who are rendered as subjects and subjected to relations of power through discourse.

3 Prior to the rise of neoliberalism, Karl Polanyi had, in *The Great Transformation* (1944), critiqued the earlier endeavour of economic liberalism to establish a self-regulating market economy. Polanyi theorized that this approach of industrial capitalism would transform humans, nature, and money into commodities. His research showed that prior to the creation of new market institutions associated with industrialization, societies based their economies on reciprocity and redistribution, but after the "great transformation" they were moulded to fit the new market-based economic institutions.

4 A Complex Adaptive System (CAS) is a group of systems that exhibit multiple interactions and feedback mechanisms in a non-linear manner to form a complex whole that has the capacity to adapt in a changing environment (Levin 1999; Gunderson and Holling 2002; Holland 2006). An understanding of the changes that CASs undergo through adaptive cycles provides insight into how to manage a system's resilience—the amount of disturbance that can be absorbed by a CAS without altering its basic structure and function (Holling 1973, 1986; Walker et al. 2004; Walker and Salt 2006). Complexity theory also adds the focus of understanding how disturbance and the timing of actions can lead to transformational social change (Westley et al. 2007).

5 The Ontario Ministry of Natural Resources was renamed the Ontario Ministry of Natural Resources and Forestry following the provincial election in June 2014.

6 For a detailed discussion about this program and other factors that contributed to its failure see Harvey (Chapter 3).

7 Staples are raw or unfinished bulk commodity products sold in export markets with minimal amounts of processing, as is the case for most Canadian forest products

(Howlett and Brownsey 2008). Staples theory describes several phases in the development of a resource staple economy where decline or crisis following depletion of the resource, rising costs, and industrial subsidization (Clapp 1998) is comparable to system collapse explained by complexity theory (Gunderson and Holling 2002).

8 See Casimirri and Kant (Chapter 4) for a case study about one northern Ontario First Nation's attempt to negotiate for recognition of its rights in relation to forest management in its traditional territory to rectify its historical marginalization from the forestry system.

9 *Tsilhqot'in Nation v. British Columbia*, 2014 SCC 44.

References

Agrawal, A., and E. Ostrom. 2001. "Collective Action, Property Rights, and Decentralization in Resource Use in India and Nepal." *Politics and Society* 29: 485–514.

Anderson, M. 2014. "Tsilhqot'in Nation Gives Canada a Chance to Do It Right: Why Our Era of Resource Giveaways May Be Over." *The Tyee*, 30 June. Accessed 13 August 2014. http://thetyee.ca/Opinion/2014/06/30/Tsilhqotin-Nation-New-Chance/.

Beinhocker, E.D. 2006. *The Origin of Wealth: Evolution, Complexity and the Radical Remaking of Economics.* Boston: Harvard Business School Press.

Bogdanski, B.E.C. 2008. *Canada's Boreal Forest Economy: Economic and Socioeconomic Issues and Research Opportunities.* Information Report BC-X-414. Victoria, BC: Natural Resources Canada, Canadian Forest Service, Pacific Forestry Centre. Accessed 21 December 2010. http://dsp-psd.pwgsc.gc.ca/collection_2008/nrcan/Fo143-2-414E.pdf.

Brenner, N., and N. Theodore. 2002. *Spaces of Neoliberalism: Urban Restructuring in North America and Europe.* Oxford: Blackwell.

Brown, S.E. 2007. "In Control: Charter Proposes Methods to Save Trees and Industry." *Chronicle Journal*, 18 September.

Bullock, R., and J. Lawler. 2014. *Community Forests Canada: Bridging Practice, Research and Advocacy.* Workshop and Symposium Report. Centre for Forest Interdisciplinary Research and Department of Environmental Studies and Sciences, Winnipeg, MB: University of Winnipeg.

Burton, P.J., C. Messier, G.F. Weetman, E.E. Prepas, W.L. Adamowicz, and R. Tittler. 2003. "The Current State of Boreal Forestry and the Drive for Change." In *Towards Sustainable Management of the Boreal Forest*, edited by P.J. Burton, C. Messier, D.W. Smith and W.L. Adamowicz, 1–40. Ottawa: NRC Research Press.

Cassidy, J. 2010. "Letter From Chicago: After the Blowup: Laissez-faire Economists Do Some Soul-Searching—and Finger-Pointing." *New Yorker,* 11 January. Accessed 25 January 2015. http://www.newyorker.com/magazine/2010/01/11/after-the-blowup.

Cavanagh, J., and R. Broad. 2012. "It's the New Economy, Stupid." *Nation* 295 (25): 18–23.

Chesters G. and I. Welsh. 2005. "Complexity and Social Movement(s): Process and Emergence in Planetary Action Systems." *Theory, Culture and Society* 22 (5): 187–211.

Clapp, R.A. 1998. "The Resource Cycle in Forestry and Fishing." *Canadian Geographer* 42 (2): 129–44.

Clark, A., and J. Treanor 2008. "Greenspan – I Was Wrong about the Economy. Sort of." *Guardian,* 24 October. Accessed 25 January 2015. http://www.theguardian.com/business/2008/oct/24/economics-creditcrunch-federal-reserve-greenspan.

Conyers, D. 1983. "Decentralization: The Latest Fashion in Development Administration?" *Public Administration and Development* 3: 97–109.

Crawford, T. 2014. "More Than a Thousand Protesters Rally against Northern Gateway Pipeline in Vancouver." *Vancouver Sun*, 10 May. Accessed 20 June 2014. http://www.vancouversun.com/news/More+than+thousand+protesters+rally+against+Northern+Gateway+pipeline+Vancouver/9827485/story.html.

DeFilippis, J., R. Fisher, and E. Shragge. 2010. *Contesting Community: The Limits and Potential of Local Organizing*. New Brunswick, NJ: Rutgers University Press.

De Landa, M. 2006. *A New Philosophy of Society: Assemblage Theory and Social Complexity*. London: Continuum.

Della Porta, D., and M. Diani. 2006. *Social Movements: An Introduction*. 2nd ed. Oxford: Blackwell.

Diani, M. 2011. "Social Movements and Collective Action." In *The Sage Handbook of Social Network Analysis*, edited by P. Carrington and J. Scott, 223–35. London: Sage.

[EA Board] Environmental Assessment Board. 1994. *Decision and Reasons for Decision on Class Environmental Assessment for Timber Management on Crown Lands in Ontario.* Toronto: Ontario Ministry of Environment. Accessed 9 June 2014. http://www.mnr.gov.on.ca/en/Business/Forests/2ColumnSubPage/STEL02_179249.html.

Engler, E. 2007. "Defining the Anti-Globalization Movement." *Encyclopedia of Activism and Social Justice*. Accessed 20 June 2014. http://www.democracyuprising.com/2007/04/anti-globalization-movement/.

Escobar, A. 2008. *Territories of Difference: Place, Movements, Life, Redes.* Durham, NC: Duke University Press.

Ferge, Z. 2000. "What Are the State Functions Neoliberalism Wants to Get Rid of? In *Not for Sale: In Defense of Public Goods,* edited by A. Anton, M. Fiske and N. Holstrom, 181–204. Boulder, CO: Westview Press.

Foucault, M. 1982. "The Subject and Power." In *Politics, Philosophy, Culture: Interviews and Other Writings 1977–1984*, edited by H.L. Dreyfus and P. Rabinow, 17–46. London: Routledge.

Freire, P. 1974. *Education for Critical Consciousness.* New York: Continuum.

Gallagher, B. 2012. *Resource Rulers: Fortune and Folly on Canada's Road to Resources.* Waterloo, ON: Bill Gallagher.

Gibson-Graham, J.K. 1996. *The End of Capitalism (As We Knew It): A Feminist Critique of Political Economy*. Oxford: Blackwell.

———. 2005. "Surplus Possibilities: Postdevelopment and Community Economies." *Singapore Journal of Tropical Geography* 26 (1): 4–26.

———. 2006. *A Postcapitalist Politics*. Minneapolis: University of Minnesota Press.

———. 2008. "Diverse Economies: Performative Practices for 'Other Worlds.'" *Progress in Human Geography* 32 (5): 613–32.

Gunderson, L.H., and C.S. Holling, eds. 2002. *Panarchy: Understanding Transformations in Human and Natural Systems*. Washington, DC: Island Press.

Harden-Donahue, A. 2014. "Reflections on the People's Climate March." The Council of Canadians, Andrea Harden-Donahue's Blog. Accessed 26 January 2015. http://www.canadians.org/blog/reflections-peoples-climate-march.

Harvey, J.T. 2010. "Neoliberalism, Neoclassicism and Economic Welfare." *Journal of Economic Issues* 44 (2): 359–67.

Hildebrandt, A. 2014. "Supreme Court's Tsilhqot'in First Nation Ruling a Game Changer for All: A Case of 'National Importance' Empowers First Nations, but May Complicate Big Resource Projects." *CBC News*, 27 July. Accessed 11 August 2014. http://www.cbc.ca/news/aboriginal/supreme-court-s-tsilhqot-in-first-nation-ruling-a-game-changer-for-all-1.2689140.

Holland, J.H. 2006. "Studying Complex Adaptive Systems." *Journal of Systems Science and Complexity* 19 (1): 1–8.

Holling, C.S. 1973. "Resilience and Stability of Ecological Systems." *Annual Review of Ecology and Systematics* 4: 1–23.

———. 1986. "The Resilience of Terrestrial Ecosystems: Local Surprise and Global Change." In *Sustainable Development of the Biosphere*, edited by W.C. Clark and R.E. Munn, 292–317. Cambridge: Cambridge University Press.

Howlett, M., and K. Brownsey. 2008. "Introduction: Toward a Post-staples State?" In *Canada's Resource Economy in Transition: The Past, Present, and Future of Canadian Staples Industries*, edited by M. Howlett and K. Brownsey, 3–15. Toronto: Emond Montgomery.

Jarvis, B. 2013. "Idle No More: Native-led Protest Movement Takes on Canadian Government." *Rolling Stone*, 4 February. Accessed 20 June 2014. http://www.rollingstone.com/politics/news/idle-no-more-native-led-protest-movement-takes-on-canadian-government-20130204.

Larson, A. 2005. "Formal Decentralization and the Imperative of Decentralization 'From Below': A Case Study of Natural Resource Management in Nicaragua." In *Democratic Decentralization through a Natural Resource Lens*, edited by J. Ribot and A. Larson, 55–70. London: Routledge.

Larson, A., and F. Soto. 2008. "Decentralization of Natural Resource Governance Regimes." *Annual Review of Environment and Resources* 33: 213–39.

Levin, S.A. 1999. *Fragile Domain: Complexity and the Commons*. Reading, UK: Perseus Books.

Leyshon, A., and R. Lee. 2003. "Conclusions: Re-making Geographies and the Construction of Spaces of Hope." In *Alternative Economic Spaces: An Introduction*, edited by A. Leyshon, R. Lee, and C. Williams, 1–26. London: Sage.

Loney, H. 2013. "Line 9 Pipeline Protests Ramp up ahead of National Energy Board Hearings." *Global News*, 8 October. Accessed 11 June 2014. http://globalnews.ca/news/887935/line-9-pipeline-protests-ramp-up-ahead-of-national-energy-board-hearings/.

McCarthy, J. 2006a. "Rural Geography: Alternative Rural Economies: The Search for Alterity in Forests, Fisheries, Food, and Fair Trade." *Progress in Human Geography* 30 (6): 803–11.

———. 2006b. "Neoliberalism and the Politics of Alternatives: Community Forestry in British Columbia and the United States." *Annals of the Association of American Geographers* 96 (1): 84–104.

McFarlane, C. 2009. "Translocal Assemblages: Space, Power and Social Movements." *Geoforum* 40 (4): 561–567.

Mirowski, P. 2014. *Never Let a Serious Crisis Go to Waste: How Neoliberalism Survived the Financial Meltdown.* Brooklyn: Verso.

Mutersbaugh, T., D. Klooster, M.-C. Renaud, and P. Taylor. 2005. "Certifying Rural Spaces: Quality-certified Products and Rural Governance." *Journal of Rural Studies* 21 (4): 381–88.

[NOSCP] Northern Ontario Sustainable Communities Partnership. 2007a. "Community-based Forest Management for Northern Ontario: A Discussion and Background Paper." Accessed 2 May 2014. http://noscp.ca/?page_id=145.

———. 2007b. "The Charter." Accessed 2 May 2014. http://noscp.ca/?page_id=37/.

———. 2009. 4–5 March Workshop Press Release. Accessed 2 May 2014. http://noscp.ca/?page_id=253.

———. 2010. "Response to Ontario's Proposed Framework to Modernize Ontario's Forest Tenure and Pricing System." Submitted to Ontario Ministry of Northern Development, Mines and Forestry, 18 May. http://noscp.ca/wp-content/uploads/2012/09/NOSCP_Charter_Response-to-tenure-consultation_18may10v2.pdf.

———. 2011. "Ontario's Forest Tenure Modernization Act: A Timid Beginning with Tons of Potential." Press Release, 20 May. Accessed 2 May 2014. http://noscp.ca/wp-content/uploads/2012/09/NOSCP_Press_Release_19may111.pdf.

Novkovic, S., and T. Webb. 2014. *Co-operatives in a Post-growth Era: Creating Co-operative Economics.* London: Zed.

[OFC] Ontario's Forestry Coalition. n.d. Sawmill Closures (Temporary and Permanent). Toronto, ON: Ontario Forest Industries Association Accessed 21 November 2013. http://www.forestrycoalition.com/closures.html.

[OFIA, NOACC, NOMA & FONOM] Ontario Forest Industries Association, Northwestern Ontario Association of Chambers of Commerce, Northwestern Ontario Municipal Association and Federation of Northern Ontario Municipalities. 2011. Letter Addressed to Party Leader, 5 July. Accessed 2 May 2014. http://www.ofia.com/files/Pre-election%20template%20letter_OFIA-NOMA-NOACC-FONOM_%20to%20Provincal%20Party%20Leaders%20of%20the%20Liberal%20NDP%20and%20PC%20parties%20July%205%202011.pdf.

[OMNDMF] Ontario Ministry of Northern Development Mines and Forestry. 2010. *Ontario's Forests, Ontario's Future: A Proposed Framework to Modernize Ontario's Forest Tenure and Pricing System.* Toronto: Queen's Printer for Ontario. Accessed 2 May 2014. http://www.mnr.gov.on.ca/stdprodconsume/groups/lr/@mnr/@forests/documents/document/stdprod_092435.pdf.

———. 2011. *Strengthening Forestry's Future: Forest Tenure Modernization in Ontario.* Toronto: Queen's Printer for Ontario. Accessed May 7, 2012. http://www.mnr.gov.on.ca/en/Business/Forests/2ColumnSubPage/STDPROD_092054.html. Also available at http://wbn.scholarsportal.info/node/10096.

[OMNR] Ontario Ministry of Natural Resources. 1998. *Toward the Development of Resource Tenure Principles in Ontario. A Discussion Paper on Natural Resource Tenure.* Toronto: Queen's Printer for Ontario.

Palmer, L., P. Smith, and C. Shahi. 2012. *Building Resilient Northern Ontario Communities Through Community-Based Forest Management.* Report from SSHRC Public Outreach Workshop 17 May 2011. Accessed 7 May 2012. http://noscp.ca/wp-content/uploads/2012/09/Workshop_Report_Final_27Jan2012.pdf.

Palmer, L., P. Smith, and R. Bullock. 2013. "Community Forests Canada: A New National Network." *Forestry Chronicle* 89 (2): 133–34.

Palmer, L., M.A. Smith, and C. Shahi. 2016. "Community Forestry on Crown Land in Northern Ontario: Emerging Paradigm or Localized Anomaly?" In *Community Forestry in Canada: Drawing Lessons from Policy and Practice*, edited by S. Teitelbaum, 94–135. Vancouver: University of British Columbia Press.

Patriquin, M.N., J.R. Parkins, and R.C. Stedman. 2009. "Bringing Home the Bacon: Industry, Employment and Income in Boreal Canada." *Forestry Chronicle* 85 (1): 65–74.

Pinkerton, E., R. Heaslip, and J.J. Silver. 2008. "Finding 'Space' for Comanagement of Forests within the Neoliberal Paradigm: Rights, Strategies, and Tools for Asserting a Local Agenda." *Human Ecology* 36: 343–55.

Polanyi, K. 1944. *The Great Transformation: The Political Origins of Our Time*. Boston: Beacon Press.

Posner, R. 2009. *A Failure of Capitalism: The Crisis of '08 and the Descent into Depression.* Cambridge, MA: Harvard University Press.

Rankin, K.N., and J. Delaney. 2011. "Community BIAs as Practices of Assemblage: Contingent Politics in the Neoliberal City." *Environment and Planning A* 43 (6): 1363–1380.

Rehmann, J. 2012. "Occupy Wall Street and the Question of Hegemony: A Gramscian Analysis." *Socialism and Democracy* 27 (1): 1–18.

Reisch, M. 2013. "Radical Community Organizing." In *The Handbook of Community Practice*, edited by M. Weil, S. Michael, and M. Reisch, 361–81. Thousand Oaks, CA: Sage.

Ribot, J. 2002. *Democratic Decentralization of Natural Resources: Institutionalizing Popular Participation.* Washington, DC: World Resources Institute.

Ribot, J., A. Agrawal, and A.M. Larson. 2006. "Recentralizing while Decentralizing: How National Governments Reappropriate Forest Resources." *World Development* 34 (11): 1864–86.

Rubin, H.J., and I.S. Rubin. 1992. *Community Organizing and Community Development.* 2nd ed. New York: Macmillan.

Ruiz Leotaud, V. 2014. "Anti-pipeline Protesters March through Downtown Vancouver." *Vancouver Observer*, 1 May. Accessed 20 June 2014. http://www.vancouverobserver.com/news/anti-pipeline-protesters-march-through-downtown-vancouver.

Schwartz, D., and M. Gollom. 2013. "New Brunswick Fracking Protests and the Fight for Aboriginal Rights: Duty to Consult at Core of Conflict over Shale Gas Development." *CBC News*, 19 October. Accessed 20 June 2014. http://www.cbc.ca/news/canada/n-b-fracking-protests-and-the-fight-for-aboriginal-rights-1.2126515.

Sharlet, J. 2013. "Inside Occupy Wall Street: How a Bunch of Anarchists and Radicals with Nothing but Sleeping Bags Launched a Nationwide Movement." *Rolling Stone*, 24 November. Accessed 20 June 2014. http://www.rollingstone.com/politics/news/occupy-wall-street-welcome-to-the-occupation-20111110.

Shragge, E. 2013. *Activism and Social Change: Lessons for Community Organizing*. 2nd ed. Toronto: University of Toronto Press.

Smith, M. 2014. "The Role of Social Movements and Interest Groups." In *Publicity and the Canadian State*, edited by K. Kozolanka, 262–80. Toronto: University of Toronto Press.

Smith, P. 1998. "Aboriginal and Treaty Rights and Aboriginal Participation: Essential Elements of Sustainable Forest Management." *Forestry Chronicle* 74 (3): 327–33.

Southcott, C. 2006. *The North in Numbers: A Demographic Analysis of Social and Economic Change in Northern Ontario*. Thunder Bay, ON: Northern Studies Press.

Speers, M. 2010. "Ontario Ministry of Northern Development, Mines and Forestry Tenure and Pricing Review." Presentation at 42nd Annual Lakehead University Forestry Symposium: A New Approach: Tenure and Reform in Ontario's Forests, held 15 January, Lakehead University, Thunder Bay, ON.

Springer, S. 2012. "Neoliberalism as Discourse: Between Foucauldian Political Economy and Marxian Poststructuralism." *Critical Discourse Studies* 9 (2): 133–47.

Staggenborg, S. 2011. *Social Movements*. 2nd ed. Toronto: Oxford University Press.

Taylor, P.L. 2005. "In the Market But Not of It: Fair Trade Coffee and Forest Stewardship Council Certification as Market-Based Social Change." *World Development* 33 (1): 129–47.

Taylor, P.L., P. Cronkleton, and D. Barry. 2013. "Learning in the Field: Using Community Self Studies to Strengthen Forest-Based Social Movements." *Sustainable Development* 21: 209–23.

Thorpe, J., and A. Sandberg. 2008. "Knotty Tales: Forest Policy Narratives in an Era of Transition." In *Canada's Resource Economy in Transition: The Past, Present, and Future of Canadian Staples Industries*, edited by M. Howlett, and K. Brownsey, 189–207. Toronto: Emond Montgomery.

Van der Veen, M. 2013. "Contending Theories of the Current Economic Crisis." *Socialism and Democracy* 27 (3): 32–53.

Walker, B., C.S. Holling, S.R. Carpenter, and A. Kinzig. 2004. "Resilience, Adaptability and Transformability in Social-ecological Systems." *Ecology and Society* 9 (2): 5. http://www.ecologyandsociety.org/vol9/iss2/art5.

Walker, B., and D. Salt. 2006. *Resilience Thinking*. Washington, DC: Island Press.

Westley, F., B. Zimmerman, and M.Q. Patton. 2007. *Getting to Maybe: How the World Is Changed*. Toronto: Vintage Canada.

[WSF] World Social Forum. 2015. Wikipedia. Accessed 25 January 2015. http://en.wikipedia.org/wiki/World_Social_Forum.

Thirty Years of Community Forestry in Ontario: Bridging the Gap Between Communities and Forestry

Stephen Harvey[1]

Communities, forests, and forestry have had a long and varied relationship in Ontario. This chapter examines the evolution of that relationship and its possible future, and provides a review of key shifts in provincial government policy affecting forestry and the course of development in northern Ontario since 1970. It highlights various milestones represented by the introduction of new policies, planning documents, and committee decisions, which have shaped the changing course of, and relationships among, forests, communities, and the practice of forestry in northern Ontario, and it discusses recent changes to Ontario's forest tenure system and new models for community involvement in forestry. The analysis points to several possible policy gaps by recounting the evolution of northern development and forestry policy in Ontario, and frames these gaps as recommendations for practice in the final section of the chapter.

The "forest" addressed is that which lies within Ontario's "Area of the Undertaking" (AOU), which is the publicly owned 27 million hectares of Crown forest in northern Ontario in which forest management activities are conducted under the direction of two key pieces of Ontario legislation, the Crown Forest Sustainability Act, 1994, and the Environmental Assessment Act, 1990. The "forest" also includes the recently added (as of 2009) 1.2-million-hectare Whitefeather Forest within the traditional territory of the people of Pikangikum First Nation. The southern boundary of the forest lies near the southern edge of the Precambrian Shield. "Community," as it is used in this chapter, refers to the geographic place where people live and work and have a form of organized local government. Some 850,000 people live in various cities, towns, and Indigenous communities in the forest. Some of the municipalities are single-industry resource-dependent communities that largely owe their existence to the forest industry, while First Nations have traditionally used the forest for a mix of self-sustaining activities, including wild foods and fur harvesting, and small-scale logging. The forest is home to more than 144 municipalities, 106 First Nations, and other Metis communities, and more than 150 unincorporated communities. "Forestry" includes the activities of planning, access road development, harvesting, renewal, and protection of the forest as well as the goods and services it produces. The goal for Ontario's forests, as stated in the Policy Framework for Sustainable Forests (1994a: 1) "is to ensure the long-term health of forest ecosystems for the benefit of the local and global environments, while enabling present and future generations to meet their material and social needs."

The Ontario Ministry of Natural Resources is responsible for developing and administering policy and law that directs forestry. As well, the ministry oversees programs of forest science and information, and forest health and protection. Many of the operational activities of forestry, including local plan development, are undertaken by the private sector as a condition of holding long-term Sustainable Forest Licences for harvest. Until recently the majority of these private operational forest management companies had a direct tie to a wood processing industry through whole or co-operative ownership; recently efforts have been undertaken to encourage the establishment of new corporate entities with less direct ties to the forest industry. This is in part an attempt to make forest management decision making more open and inclusive, and to provide opportunities for new entrants, as well as strengthen the role of communities in the planning and development of the forest.

Communities and Forestry in Ontario's Past

Forestry and associated settlement began in the southern portion of what is now known as the "Area of the Undertaking" by the mid 1850s as available lands for settlement were taken up and the forests in agricultural Ontario were depleted. In some areas forestry complemented farming, with farmers working in the forest in the winter and returning to their farms in the spring. In other areas, especially those not suitable for agriculture, temporary communities were established around lumber mills and mills were moved or closed in tune with their economic fortunes. The arrival of the twentieth century saw larger communities emerge to service both the pulp and paper and lumber industries; some of these were expansions of existing settlements and others were new towns. In the more remote "northern Ontario," a number of the new towns were "company towns" planned solely for the purpose of housing and supporting workers in the mills. These towns were in the northern "hinterland" of the province, sometimes located large distances from the centre of population in southern Ontario (for further description of such towns see Casimirri and Kant, Chapter 4; Lachance, Chapter 5; and Palmer and Smith, Chapter 2). Ontario prospered and grew, and northern Ontario municipalities, though to a lesser extent, grew as well. By 1900, the sale of timber accounted for as much as one third of the province's direct revenues (Armson 2001). Generally speaking, although not necessarily the case with Indigenous communities, community development objectives were closely aligned with those of the forest sector, or at least they appeared to be as long as both were growing. What was good for the mills was good for communities.

The province's plan for northern development was simple: make land available for settlement and draw investors with access to natural resources. This strategy worked well for the growth of many non-Indigenous communities for almost eighty years. While there are some stark examples of individual failures and hardships, the municipalities of northern Ontario continued to grow until the 1970s. For the past thirty years the population of the region has stagnated, even showing a slight decline over the past ten years (Southcott 2013).

Forestry has also seen many changes as it moved through an initial period focused on timber exploitation, to a second period focused on administration and orderly development, and finally to the current era which is focused on the drive to achieve sustainable forestry. Until the advent of sustainable forestry, and its more holistic view of forests ecosystems and sustainable communities, there was no explicit government policy connecting forests and communities. Government took its role in administering forestry very seriously and built a large and robust system and accompanying bureaucracy charged with looking after the forest, forest lands, fish, and wildlife for all the people of Ontario. Until recently, communities in the forest, for the most part, did not intervene or attempt to define a role for themselves in forest management. Instead, communities administered local planning and municipal affairs within their legislatively defined geographic boundaries. The province alone looked after Crown forest lands, and cities and towns provided services and supported development on municipal lands.

The politics of forestry began to change in the 1960s, driven by the emergence of several new groups and government entities focused on environmental and development issues. For example, public outcry over logging in Algonquin

Park resulted in the establishment of a public advocacy group—the Algonquin Wildlands League (now Wildlands League, a chapter of the Canadian Parks and Wilderness Society), which concerned itself initially with that matter alone. This public attention to forestry in Ontario's showcase park was enough for the government to reduce the extent of logging in Algonquin Park. In 1974, a public (Crown) agency, the Algonquin Forestry Authority, governed by appointed citizens mostly from surrounding local communities, was also created to oversee forestry operations in the park.

In the mid-1970s the Ontario government created a stand-alone Ministry of Northern Affairs supported by an Act of the same name. As the only department with a regional mandate, this ministry and the legislation were a response to those who asserted that government was not listening to resource-dependent and physically remote communities. Later the government changed the name to the Ministry of Northern Development and then added Mines (MNDM) to reflect a mandate emphasizing development. In 2009, forestry was added to MNDM, for a short period, as part of an overall commitment to address the economic crisis in the forest sector that was underway at the time.

The establishment of the Ministry of Northern Affairs was an early attempt to reassure northern Ontario citizens that their needs were being attended to and that there was an opportunity to influence government decision making, including decisions related to forestry. Soon after its creation, in 1977 the Ontario Royal Commission on the Northern Environment (RCNE) was struck to examine the challenges facing the far north (i.e., the area north of 50 degrees latitude), including its peoples, communities, and natural environment. After years of research and public hearings the commission reported back to government in 1985 with a comprehensive, but lacklustre, report and recommendations (see Fahlgren 1985). Among the many recommendations was the identified need to establish a Northern Development Authority. In spite of its struggles to furnish the government with a clear plan for orderly development in the north, the RCNE stands as further evidence that the government was aware of its unique role in northern affairs. A change in government and the delay in receiving the commission's final report meant that the report did not have a significant impact as government turned its attention to other matters.

Community groups also emerged and contributed to discussions related to environment and development in northern Ontario. For example, the Lakehead Social Planning Council (LSPC) was established in 1963 in Thunder Bay with a stated mission of "building a better community through strategic alliances, social research and the provision of valid, reliable information" (LSPC

2012). In 1981 the council published a study examining the future of the forest industry in northwestern Ontario. This interest by a community-based organization, alongside the adage "what was good for the mill was good for the community," provides an early indication that the century-old relationship between communities and the forest sector was ripe for change. At the same time as the RCNE was reporting, the government, under different political leadership, took a very pragmatic approach and struck the Advisory Committee on Resource Dependent Communities, comprised of elected members of provincial parliament and citizens, to further study the problems of resource-dependent communities and recommend ways for government to address the unique needs of this distinct, and very large but sparsely populated, geographic region. In pursuing the direction from the Minister of Northern Development and Mines, the Advisory Committee on Resource Dependent Communities reported back in 1986 with a host of recommendations including a request for government to initiate sectoral planning for forestry (see Rosehart 1986). Both the LSCP and the committee pointed to the need to strengthen the link between community development and forest sector development planning.

The next major step in northern planning and policy came in 1989, when Larry Taman, assistant deputy minister of the Ministry of the Attorney General, and a highly regarded legal advisor, led an inter-ministerial task force on northern governance. The task force, in an internal memo, noted the following:[2]

1. the legitimacy of government policy in the north is often challenged because northern residents lack the self-government, autonomy and responsibility of southern residents (afforded by the complete coverage of the south by municipal governments);
2. the need for devolution of responsibilities has been advocated by numerous government studies; empowerment has been a theme of various governments yet there has been no coherent northern initiative by government as a whole in this regard;
3. the commitment to resolve Aboriginal land claims and self-government aspirations contrasts with the lack of comparable commitment to self-government for other northern residents;
4. although direct provincial administration of the north is, strictly speaking, politically accountable to the people there, in practice it is government by provincial civil servants, who are typically based in southern Ontario.

These observations indicated senior government's growing awareness of the desires of northern communities and governments to obtain more control over their own geographically unique planning issues, and the need to develop government policy and decision making that was more representative and accountable to northern residents. In part, these policy events, which made explicit the direct links between local socio-economic issues, decision making, and natural resource development, set the stage for community-based resource management in northern Ontario.

Throughout the 1990s the Conservation Council of Ontario and the community of Geraldton, located northeast of Thunder Bay, discussed community forestry as an alternative to the status quo. Regular contact among Canadian Forest Service, town council, and Conservation Council members developed a mutual appetite for innovation and collaboration. At the same time, in the late 1990s the Class Environmental Assessment of Timber Management on Crown Lands in Ontario, the work of Ontario's Forest Policy Panel, and the 1994 Crown Forest Sustainability Act (CFSA) institutionalized public involvement in forest policy and planning throughout the province. It seemed that institutions and ideas were shifting in favour of public and community involvement in forest development.

In an effort to redefine the course of forestry, the Ontario government announced a sustainable forestry program in 1991 and its cornerstone, the development of a comprehensive forest policy framework. Ontario's Policy Framework for Sustainable Forests (OMNR 1994a), a statement of strategic intent and direction, set out principles for decision making, emphasizing the need for adaptive management. The framework's goal, principles, and objectives were developed by a panel of appointed "expert citizens" who held public meetings and engaged in dialogue with over 3,000 people to compile findings. Their report, *Diversity: Forests, People, Communities: A Comprehensive Forest Policy Framework for Ontario* (OFPP 1993), served as the foundation for the CFSA. A key component of Ontario's sustainable forestry program was a community forestry project with its goal of devising a strategy to enhance opportunities for "local participative management in forestry." The community forestry project was a significant commitment by Ontario and some communities to explore the concept of community forestry and the relationships between communities and the extensive area of Crown forest in northern Ontario. Four community forest pilot projects were launched in 1992 as a partnership between communities and the Ontario Ministry of Natural Resources.

In addition to sponsoring four community forestry pilot projects, the provincial initiative also undertook a study of the institutional arrangements which set out a role for communities in natural resource management. While the focus of the study *Partnerships for Community Involvement in Forestry* (OMNR 1994b) was Ontario, efforts were made to examine relevant institutions in the United States and British Columbia. Based on this work and the advice of a team of planning consultants, a set of requirements for a successful community forest was proposed (as originally published by the author in Harvey and Hillier 1994: 729):

1. **Community consensus.** There should be a fair degree of consensus within the partner community around environmental and resource management values, and in particular the kinds of objectives and expectations that would underlie a community forestry partnership. Such a consensus should already be present in the community; it cannot be made to happen.
2. **Credible lead agency**. The lead agency should have high credibility throughout the partner community, a low profile in the partner community, or a combination of these that will ensure that its participation is accepted and trusted.
3. **Clear mission**. The partnership should have a clearly articulated mission to achieve explicit economic, employment, social, and/or cultural benefits as desired by the partner community.
4. **Meaningful delegation**. There should be a significant and meaningful delegation of authority and responsibility from the lead agency to the partnership.
5. **Meaningful tenure**. The partnership should be granted some form of meaningful forest tenure, long and secure enough to impart some sense of co-ownership, and involving a delegation of responsibility consistent with the tenure type.
6. **Meaningful revenue autonomy**. Whether through its forest tenure, other structural arrangements, or some combination of these, the partnership should be enabled with sufficient revenue sources and autonomy that it can effectively achieve its objectives and maintain a high degree of financial independence.
7. **Meaningful inclusion of interests**. All relevant interests should be represented in the partnership, or should otherwise be sufficiently included in the decision-making process that they will receive and appear to receive fair consideration.

8. **Reliance on existing structures**. Wherever possible, the partnership should incorporate an existing organized community—normally a municipality or Indian band—and rely on that community's structures where at all possible. The partnership should fit within a standardized decision-making, regulatory, and procedural framework for resource management, whether an existing tenure type, or a new framework developed for community forestry.

With a deepening recession that eventually served to distract government from most business other than fiscal matters, a community forestry strategy was never completed, leaving the province without a common set of objectives or policy for communities, forests, and the forest sector. The community forestry pilot project, however, was not without positive results. The CFSA contained a provision for the establishment of Forest Management Boards for "areas as are designated by the Minister, including forest management boards for community forests designated by the Minister" (Section 15). This provision opened the door to direct local involvement in forestry decision making.

In 1995, a new government emphasized the need for a smaller government and reduced expenditures. Efforts were launched to privatize the delivery of those forest services that remained in the hands of the Ministry of Natural Resources. With the valuable learning experience of Ontario's community forestry project in hand, Ontario's only community-based non-profit forest management entity, Westwind Forest Stewardship Inc., was established. While Westwind was established ostensibly to address the government's goal of privatizing forest operations and downsizing government, local leaders seized the opportunity to create an effective opportunity for citizen involvement, and Westwind was incorporated in 1997. Several years after it was formed, Westwind, and the forest it managed, became the first Forest Stewardship Council (FSC) certified public forest in Canada.

The lessons learned from Ontario's community forestry project are still relevant today. From a community planning perspective the government's interest in community forestry was an early signal that the relationship between the forest management interests of the traditional forest industries and the interests of communities needed closer examination. The emergence of the Northern Ontario Sustainable Communities Partnership (NOSCP), a grassroots effort of First Nations, municipalities, interested academics, and community members, is evidence of a growing desire by northern Ontarians to establish greater autonomy over the natural resources in the region (see Palmer and Smith, Chapter 2).

New Paths for Communities in Ontario Forest Governance

After Westwind was created in 1997, not a great deal happened in Ontario that can be related to provincial-led community forestry. For some it was perhaps a waiting period, a time to validate opinions that community forestry would not work in Ontario. Critics of Westwind waited for the institution to fail; it did not. In fact, in 2009 when the Ontario government began to seriously question the effectiveness of its policy for forest tenure and pricing, some began to examine Westwind, not with an interest in exposing its weaknesses but, rather, to learn. Of particular interest was the "governance" structure of Westwind as an alternative model for multi-stakeholder collaboration and forestry decision making. Until this time, leaders in forest policy showed little interest in governance and its application to organizations operating somewhere between the private sector and "big government." However, the broader inclusion of non-conventional forestry actors would also necessitate tenure reforms.

The Road Ahead: What's Changing?

Forest tenure is the term commonly used to describe the allocation and licensing of timber from Crown forests. Historically, tenure is governed by legal arrangements that define the rights and responsibilities assigned to forestry companies and other resource users when they are issued a licence to access resources on publicly owned land. Companies in Ontario holding a licence for timber on Crown forests can harvest that timber only after a plan has been approved, a plan that addresses the government's requirements, as laid out in the CFSA and Ontario's Forest Management Planning Manual, for sustainable forestry. Forest management, including planning, is undertaken locally on some forty-plus forest management units that make up the "Area of the Undertaking" and the Whitefeather Forest. The Sustainable Forest Licence holder is responsible for ensuring that a forest management plan is prepared and then approved by the Ministry of Natural Resources.

Ontario has no plan to radically change the relationship between communities and forestry, and the province does not yet have institutionalized community forests. In 2011, Ontario approved a regional economic development plan (Growth Plan for Northern Ontario), but the plan does not address the relationship between communities and forestry, and provides little direction for the integration of community development planning with forestry. There is, however, some modest change aimed at enhancing community involvement in forestry. Efforts to modernize and renew Ontario's forest tenure system are underway. The change is significant and arguably one of the biggest changes to forest policy since the 1990s. The Forest Tenure Modernization Act was passed in June 2011; the Act amends the CFSA. It would be speculative to suggest a time frame to achieve the government's objectives for the overall process of modernizing tenure as enabled by the amended CFSA. The relationship between forestry and Ontario's Aboriginal peoples is receiving considerable attention as well.

Overall, communities are considered in the new government approach. For example, the second stated objective for a modernized tenure system is to "provide opportunities for meaningful involvement by local and Aboriginal communities" (KBM Resources Group 2016). Section 5 of the Forest Tenure Modernization Act (2011) states that the following are the objects of an Ontario local forest management corporation:

1. To hold forest resource licences and manage Crown forests in a manner necessary to provide for sustainability of Crown forests in accordance with the *Crown Forest Sustainability Act, 1994* and to promote the sustainability of Crown forests.
2. To provide for economic development opportunities for aboriginal peoples.

Some communities have taken a more active role in exploring the possible opportunities to redefine their relationship with forestry through a modernized forest tenure system. Plans are underway to establish two Local Forest Management Corporations (LFMCs)—Crown Agencies (similar to that of the Algonquin Forestry Authority) governed by a predominantly local board of directors responsible for managing Crown forests and overseeing the marketing and sale of the timber in a given area. In June 2012, Ontario announced the establishment of the province's first Local Forest Management Corporation, the Nawiinginokiima Forest Management Corporation, to re-energize the forestry sector, create jobs, and boost the economy. This government agency will manage and oversee the sale of timber along the northeast shore of Lake Superior. The plan calls for the corporation to manage five existing forest management units: Nagagami Forest, White River Forest, Big Pic Forest, Black River Forest, and the Pic River Ojibway Forest. Nawiinginokiima became operational in April 2013, and efforts to create the second LFMC are underway.

Including LFMCs as part of a government plan for meaningful local involvement may be seen by some as a step toward bridging the gap between communities and forestry governance in Ontario. It is, however, a cautious step at best, since the legislation requires an evaluation of the LFMCs before any more may be established. Still, this represents a significant change in forest governance and broadening of stakeholder involvement in a region that has been primarily controlled by government and industry.

For some while, Ontario has also been challenged to increase Indigenous participation in forestry (as illustrated by Casimirri and Kant, Chapter 4). Progress has and will undoubtedly be made and there are indications that the pace of change is increasing. Enhanced relationships are emerging between Indigenous communities and forestry. Part of this new relationship is enabled by the modernization of the forest tenure system, but perhaps far more telling is the fact that Indigenous communities are involved in governance of the Nawiinginokiima Forest Management Corporation: forest licences have been issued to Indigenous communities, a relatively recent phenomenon; and the Whitefeather Forest in far northwestern Ontario was added recently to the area of Crown forest under active forest management. The community of Pikangikum was intimately involved in developing the land use strategy for the Whitefeather Forest, a step required before a forest management plan would be developed. Pikangikum First Nation elders played a guiding role in forest management planning, and their involvement was recognized by an appendix added to the provincial Forest Management Planning Manual in 2006. Moreover, in 2010, Miitigoog Limited Partnership, a partnership of local forest industry and First Nations, assumed responsibility for the management of the 1.2-million-hectare Kenora Forest. And there are other areas in which Indigenous peoples are becoming more involved in forestry such as logging contracting and in negotiations for Enhanced Sustainable Forest Licences, a second new form of tenure created by the province as part of tenure reform. Each of these examples indicates that Indigenous participation in forestry in Ontario is increasing.

Future Challenges for Communities and Forestry

There are several challenges facing efforts to narrow the gap between communities and forestry management in Ontario. First, there is a need to develop some common understanding of what is meant by "community forestry," and a track record of trials and evaluations of different models would assist with this. Second, inasmuch as community forestry involves some transfer of management authority to a local community, there is a need to devise the means that will enable the "community" to effectively address local and broader public interests—a means that conveys the sense of ownership that can spawn commitment, while ensuring public accountability. Further, balancing local interests with those of the broader public will become more of a challenge as the population becomes concentrated in the urbanized southern Ontario outside of northern forests. Third, advances in bridging the gap between community and forestry must be fiscally responsible if they are to be supported by government (and further economic analyses identifying the benefits of community forestry will be required: see Robinson, Chapter 12). Finally, forests are a part of the Canadian identity, and changes to the ownership of the forest, real or perceived, will be successful only if this is addressed.

Below are several closing recommendations for bridging key policy gaps that have persisted throughout the past thirty years of forestry policy in Ontario:

1. Further definition of the concept of "community forestry" would assist the discussion. As discussed by Bullock, Broad, Palmer, and Smith in the opening chapter of this book, community forestry is a concept, term, and practice that has been interpreted in various ways. Whether or not community forestry lends itself to a precise definition, there is a pragmatic need for those proposing it as an alternative governance model to define it in terms to which others can respond. Change for the sake of change, which some community forest advocates appear to be suggesting, is not good public policy. The principles identified during the work of the OMNR's community forestry pilot initiative, and the Charter of the Northern Ontario Sustainable Communities Partnership (see Palmer and Smith, Chapter 2), are helpful in framing a discussion of community forestry in Ontario.
2. If community forestry is to become a widespread practice in Ontario, a track record and common understanding of the concept would assist with understanding its role or potential role. A track record could build confidence in those who might otherwise be skeptical of change to the status quo. Publicly accepted evaluation criteria would provide some common terms for further discussion of the implications of community forestry. There is some experience that could help in developing publicly acceptable evaluation criteria. Westwind is a "pioneer" and not without its critics, but it has endured, and it has been the subject of various evaluations. The ongoing discussion provided by the Northern Ontario Sustainable Communities Partnership can assist with the development of public consensus. A common understanding or definition of community forestry could also assist with the development of some common criteria for its evaluation. The move to establish LFMCs is accompanied by a

commitment to evaluate the experience before establishing any more; when and how this will take place is uncertain, but the outcome will inform further discussion of community involvement in forestry.

3. There is a need and opportunity to demonstrate how community forestry can assist with balancing local and broader public interests. Each local forest reflects a balance of general societal/provincial and local interests. Sometimes these interests are closely aligned and at other times they are not. Where there is a wide gap between the local and provincial interests, decisions are often delayed. Delay often leads to lost development opportunities. The challenge of finding this balance is exacerbated by a growing dichotomy between the residents of northern Ontario, who are directly affected by forestry and are often engaged in local forest management, and the overwhelming majority of the population living in the south, who often wish to see good environmental stewardship but lack the connection with the consequences of their proposals. Northern Ontario has found itself distanced from the rest of Ontario, and its development needs are vastly different from the hi-tech, manufacturing, and agrarian economies of the more densely populated south. The forest sector has rebounded somewhat from a deep recession of 2007–11, but significant challenges remain. With an uncertain economic future, northern Ontario's non-Indigenous population is decreasing in real and relative terms, and youth out-migration remains problematic (Southcott 2006) in contrast to the steadily growing population to the south. First Nation communities, on the other hand, are young and growing and there is reason to believe that they will make an increasingly important contribution to the development of the north.
4. There is a need for local governance structures that are responsible and accountable. Much of the forest is unorganized territory; there is no formal municipal structure of cities, towns, counties, and townships as exists in southern Ontario. Ontario's Municipal and Planning Acts, which provide organized municipalities with an operating mandate, directly address forestry only within municipal boundaries. Band councils governing Indian reserves have similar or perhaps less authority to consider forest management although, in practice, they may demonstrate greater interest and control. While Westwind, Nawiinginokiima Forest Management Corporation, and the Algonquin Forestry Authority all embrace some aspect of local accountability and control in their governance structures,

none is directly tied to organized municipalities and two of the three are agencies of the Ontario government. Conservation Authorities (CAs) were examined in the 1994 study *Partnerships for Community Involvement in Forestry* (OMNR 1994b). CAs are an interesting model of governance and natural resource stewardship in Ontario. They deliver watershed programs, including flood and erosion protection and other water and land conservation–related services on behalf of member municipalities. They are, by definition, made up of neighbouring municipalities within a watershed. However, CAs are limited in northern Ontario (there are only five) because much of the region lacks municipal organization.

5. Changes to the current system are necessary to inspire innovation and recognize that some risk is inherent as a part of the learning that occurs when one "empowers" while maintaining overall public confidence in the system. Community forestry suggests decentralization and redistribution of control. As the above-mentioned experiences illustrate, community forestry in Ontario is far from a well-practised, institutionalized phenomenon. Because of its developmental nature, there is an innovative aspect to community forestry, and variety in context suggests that there will need to be diversity in approaches to implementation. Overzealous pursuit of risk management, consistency, and control, a behaviour familiar to both government and non-government organizations, can stifle innovation and commitment and may also fail to support decisions that reflect both local circumstances and strategic intent.
6. Finally, community forestry advocates need to demonstrate how the approach can contribute to the government drive to balance budgets. This will be particularly challenging in a risk-averse and challenged fiscal environment where there may be a need for some initial investment if the approach includes plans for building capacity, plans which often entail considerable cost.

Notes

1 This chapter is based on thirty years of professional experience with public policy in Ontario, but the content is entirely the perspective of the author and must not be attributed to the Ontario Ministry of Natural Resources and Forestry.

2 Internal source from the author's archives.

References

Armson, K.A. 2001. *Annual Report of the Commissioner of Crown Lands and the Forests of Ontario*. Ontario, Canada.

Fahlgren, J.E.J. 1985. *The Royal Commission on the Northern Environment: Final Report and Recommendations*. Toronto: Ontario Ministry of the Attorney General.

Harvey, S., and B. Hillier. 1994. "Community Forestry in Ontario." *Forestry Chronicle* 70 (6): 725–30.

KBM Resources Group. 2016. "Review of Forest Tenure Models." Thunder Bay, Ontario.

[LSPC] Lakehead Social Planning Council. 2012. "Vision, Mission, & Values." Accessed 12 March 2015. http://www.lspc.ca/?pgid=27.

[OFPP] Ontario Forest Policy Panel. 1993. *Diversity: Forests People, Communities: A Comprehensive Forest Policy Framework for Ontario*. Toronto: Queen's Printer for Ontario.

[OMNR] Ontario Ministry of Natural Resources, Forest Policy Branch. 1994a. "Policy Framework for Sustainable Forests." Accessed 15 March 2015. https://dr6j45jk9xcmk.cloudfront.net/documents/2826/policy-framewrk-eng-aoda.pdf.

———. 1994b. *Partnerships for Community Involvement in Forestry: A Comparative Analysis of Community Involvement in Natural Resource Management*, prepared by the Community Forest Project. Ontario: Ministry of Natural Resources.

Rosehart, B. 1986. *Final Report and Recommendations of the Advisory Committee on Resource Dependent Communities in Northern Ontario*. Toronto: Ontario Ministry of Northern Development and Mines.

Southcott, C. 2006. *The North in Numbers: A Demographic Analysis of Social and Economic Change in Northern Ontario*. Thunder Bay: Lakehead University, Centre for Northern Studies.

———. 2013. "Regional Economic Development and Socio-Economic Change in Northern Ontario." In *Governance in Northern Ontario: Economic Development and Policy Making*, edited by C. Conteh and B. Segsworth, 16–42. Toronto: University of Toronto Press.

Factors Affecting Success in a First Nation, Government, and Forest Industry Collaborative Process

Giuliana Casimirri and Shashi Kant

First Nation and non–First Nation communities, governments, and forest companies in Ontario's productive forest region will need to collaborate to resolve shared social, ecological, and economic challenges to their prosperity and sustainability. The determinants of successful cross-cultural collaboration for the management of natural resources in Canada are not well documented (for exceptions see Natcher et al. 2005; Pinkerton 1989; Robinson 2010; Wyatt et al. 2010). Previous research suggests that divergent worldviews, deeply distrustful relationships, and capacity, resource, and power disparities, as well as differences in how Indigenous and non-Indigenous collaboration participants perceive their relative rights and responsibilities for forests, are likely to serve as persistent barriers to effective and equitable cross-cultural collaboration (Ebbin 2011; Natcher et al. 2005; Smith 2007; Wyatt et al. 2010). Some researchers argue that overcoming cross-cultural collaboration challenges may require attending to underlying political questions that currently limit Indigenous peoples' decision-making authority, either through policy changes, forest tenure reform, or some other institutional change (Coombes and Hill 2005; Donoghue et al. 2010; Natcher et al. 2005; Passelac-Ross and Smith 2013).

We must understand the factors that facilitate or discourage meaningful outcomes in cross-cultural collaboration settings if we are to advance alternative forest governance models such as community forests. The critical distinguishing features of community forestry identified by Bullock, Teitelbaum, and Lawler, in Chapter 1, include: enhanced local control over resource decision making; the integration of multiple values and forms of knowledge; and local distribution of diverse benefits. These are items that have been pursued by First Nation communities in northern Ontario for some time. However, there are very few examples of natural resource management collaboration involving First Nation communities that begin to approach these community forestry conditions. More often, First Nation efforts to regain autonomy over traditional territories have involved conflict, mistrust, and misunderstanding. Community forestry policy advocacy in northern Ontario would benefit from a much better understanding of how cross-cultural collaborators have experienced and responded to challenges in developing collaborative outcomes. This chapter describes several interdependent factors that acted as barriers to or facilitated positive outcomes in one Ontario boreal forest cross-cultural collaboration effort. This effort involved the Constance Lake First Nation (CLFN) and representatives from the forest industry, the Ontario Ministry of Natural Resources (OMNR), and Aboriginal Affairs and Northern Development Canada (formerly Indian and Northern Affairs Canada [INAC]).

The CLFN negotiation process, initiated in 1997 and focusing on the Hearst Sustainable Forest Licence (SFL) area (Figure 4.1), was one of the first formal sustained and comprehensive negotiation processes in Ontario. The process was intended to resolve forest management issues of concern to the First Nation community, such as access to forest resources, Aboriginal and treaty rights affected by forest harvesting, First Nation involvement in forest management planning and management, and access to training and employment. The negotiation process came about because the CLFN had protested against harvesting on traplines in their traditional territory by blocking an access road to a local lumber mill and because of a requirement for the OMNR to undertake negotiations with Aboriginal Peoples as a condition of the Timber Class Environmental Assessment (Timber EA) approval (EAB 1994). At the time of fieldwork for this chapter in 2003, formal negotiations had ended with a draft but unsigned written agreement, and the parties to the CLFN negotiation process largely considered the formal negotiation process a "failure." However, the parties continued to work collaboratively on a number of projects and identified several positive outcomes and lessons learned.

The CLFN negotiation process serves as a useful case example of a gradual shift occurring in forest management in Ontario, from the almost complete exclusion of First Nation communities and their rights, values, knowledge, and economic development needs, to their inclusion, first as stakeholders (NAFA 1995a), and increasingly as business partners (NAFA/IOG 2000) and, to some

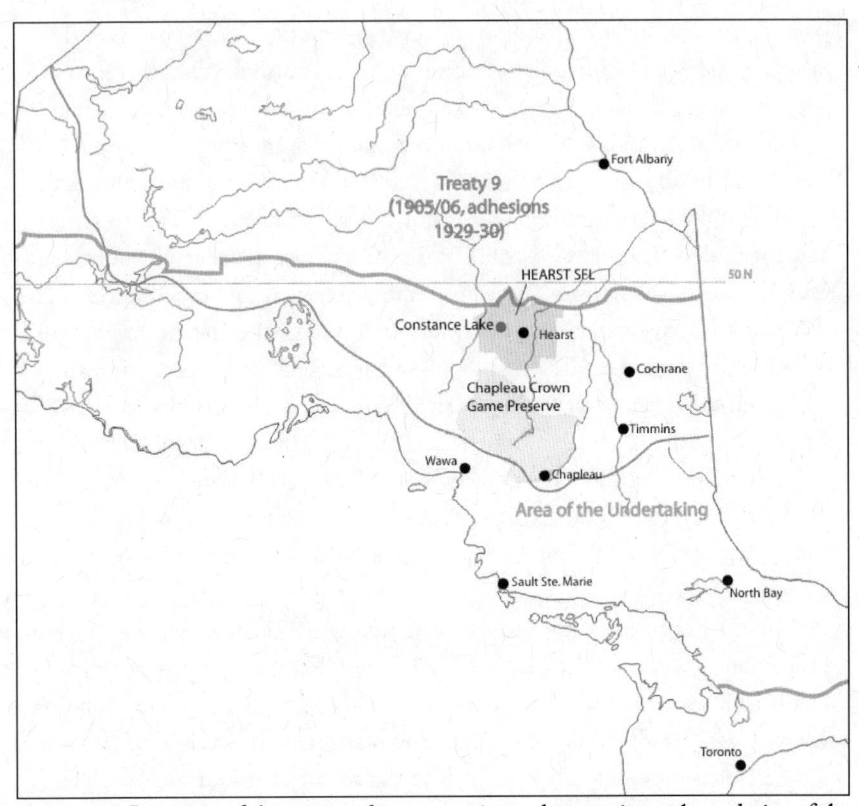

Figure 4.1. Location of the case study community and approximate boundaries of the Hearst SFL, the Treaty 9 area, and the area of the undertaking.

degree, as rights holders (Ontario 2006; Wyatt 2008). As Castro and Nielsen (2001) caution, new collaborative institutions for natural resource management, which are equitable, respectful of Aboriginal rights, and have concern for long-term sustainability, cannot be crafted without large investments of time, financial resources, social capital and, in all likelihood, many false starts. In the interim, every cross-cultural collaboration experience, regardless of its relative success or failure, has much to teach us about the procedural and contextual factors required to facilitate co-operative action, despite divergent expectations and worldviews.

The parties to the CLFN negotiation process described different understandings of the scope of the problem to be addressed, and the limitations of the parties' relationships for fostering fundamental change, as barriers to collaborative outcomes. Additional collaboration barriers that were important in this case are described elsewhere (Casimirri 2013). The above factors are highlighted in this chapter because they demonstrate that cross-cultural collaboration outcomes are more likely to be limited by contextual factors than the mechanics of the process. Participants identified the development of shared goals and joint action as positive outcomes of the process. Such outcomes were attributed to two important factors: the presence of supportive leaders and the participants' recognition of their interdependence. These factors contributed to the parties shifting to a collaborative regional economic development approach for their interactions in subsequent years. This chapter illustrates that supportive leaders who champion collaboration, mutual recognition of interdependence, development of shared goals, and acknowledgement of differing worldviews are important to cross-cultural collaboration success.

Cross-Cultural Collaboration in Natural Resource Management

Cross-cultural natural resource management collaboration refers in this chapter to collaborative processes wherein First Nation and non–First Nation representatives with cultural or worldview differences participate in a problem-solving process aimed at a "pooling of appreciations and/or tangible resources, e.g., information, money, labor, etc. to constructively explore their differences and search for solutions" (Gray 1989: 5). The term "cross-cultural collaboration" also acknowledges the cultural and worldview differences of participants, without specifying a particular institutional form, or the degree to which decision-making authority is devolved (Carlsson and Berkes 2005; Plummer and FitzGibbon 2004).

Cross-cultural collaboration and its drivers are unique in many ways. In Ontario, most First Nations are treaty partners with the Crown, and thus natural resource management collaboration efforts are inherently political. There is little shared understanding about what "collaboration" requires and what it should accomplish. For the provincial government and resource industries, cross-cultural collaboration with First Nations is undertaken to meet legal obligations, mitigate conflict, and incorporate Indigenous "values" to achieve management direction consensus. When formal collaboration efforts do occur, as in the CLFN negotiation process, they often begin with no formal mandate, no clear terms of reference, and uncertainty about each group's role and responsibilities.

The literature related to the evaluation of collaborative natural resource management has focused either on understanding the internal process characteristics under which stakeholders act collaboratively (Innes 2004; Muro and Jeffrey 2008) or on understanding the social, ecological, or political context that influences collaboration outcomes (Davis 2008; McCool and Guthrie

2001). Attention has focused on defining the attributes of a "good" collaboration process or the characteristics of "authentic dialogue." Accordingly, "good" collaboration processes must foster open and high-quality deliberation, be open to diverse participants, ensure participants set their own ground rules for behaviour, agenda setting, and decision making, and employ interest-based bargaining (IBN) to develop integrative goals (Booher and Innes 2002; Innes 2004). IBN is an approach to conflict resolution that seeks to identify the interests that underlie negotiation participants' issues and avoid predetermined bargaining positions. A collaboration effort following these "principles of good practice" (Marshall et al. 2010: 65) is expected to reduce conflict and produce better management decisions, positive social-relational outcomes, and collective action—positive outcomes commonly associated with collaboration (see Table 4.1 adapted from Casimirri 2013).

Table 4.1. Common collaborative natural resource management process-related criteria.

Process Elements
Representation • a full range of interests should be represented • the inclusion of all relevant stakeholders • stakeholders are involved as early as possible • substantial constituent outreach efforts and public involvement
Scope and Objectives • meaningful and achievable task for participants • conflict/problem/scope is mutually defined by participants • clear goals and objectives developed by the participants
Deliberation and Dialogue • mutually agreed-upon ground rules for behaviour, agenda setting, and decision-making method • two-way flow of information • communication with constituencies • open communication which permits all assumptions to be questioned • constructive conflict management • focus on joint gains, exploring interests, and reframing positions and value differences into superordinate goals

Facilitation, Information, and Resources

- skilled facilitator (preferably a neutral third party)
- full information sharing
- accessible and high-quality information
- joint fact finding and development of shared understanding
- incorporation of different kinds of knowledge
- sufficient funding for process
- collaborative leadership skills of participants
- stable staffing and participation

Sources: Abelson et al. 2003; Beckley et al. 2005; Gray 1989; Gunton and Day 2003; Innes 2004; Margerum 2011; Moore 1996; Sheppard 2005; Susskind and Cruikshank 1987.

Key among the principles of good collaboration practice is the prerequisite that collaboration participants develop clear terms of reference or a shared understanding of the problem and scope of the negotiation prior to undertaking substantive negotiations. Additionally, some scholars contend that improved relationships among partners are social capital outcomes of collaboration (Cullen et al. 2010; Wyatt et al. 2010) that can be used as measures of success to evaluate collaboration. While most evaluations of collaboration assess social outcomes, including the status of relationships among partners, in cross-cultural settings participants may not consider these "outcomes," for example, conflict reduction or enhanced personal relationships, as sufficient or even uniformly beneficial.

Other research suggests that internal conditions of the process and the dialogue are always tempered or influenced by the external conditions, for example, site and resource factors, institutional and political limitations and power asymmetries, or social and relational factors (Table 4.2 adapted from Casimirri 2013). It is also increasingly acknowledged that the capacity of those affected by a decision to deliberate in the production of that decision is influenced by the nature of the conflict and that collaboration outcomes are more likely to be determined by how well participants manage their different and often competing frames or worldviews, rather than the internal conditions of the process (Bell and Kahane 2004; Ebbin 2011; Tipa and Welch 2006).

Table 4.2. Common collaborative natural resource management context-related criteria.

Context Elements
Site and Resource Factors • resource characteristics • management setting and community characteristics • geographic scale/complexity • public awareness of the problem • degree of shared jurisdiction • adequate scientific and technological information
Institutional Factors • characteristics of the sponsoring agency or institutions • legislative, budgetary, administrative, and policy constraints/support • lack of authority and flexibility at local level • scope of collaboration limited by political directives of the respective organizations • short-term political culture
Social and Relational Factors • attitudes and skills of participants • participants' resistance to change • prior relations/perceived interdependence of participants • level of trust among participants • level of community social capital • ideological/value differences • power differences

Sources: Ansell and Gash 2007; Beierle and Konisky 2000; Bidwell and Ryan 2006; Leach and Pelky 2001; Marshall et al. 2010; Selin and Chavez 1995; Shindler et al. 1999; Wondolleck and Yaffee 2000.

The limited literature on cross-cultural collaboration has demonstrated that the development of collective action is possible when parties develop a shared ideology (Waage 2001), construct a place-based identity frame (Docherty 2004), and include "cultural mediators" who can communicate across worldviews (Ebbin 2011). Research also suggests that where cross-cultural collaboration is sustained for long time periods and supported by an enabling policy context it contributes to positive social and ecological outcomes (Armitage et al. 2011). Despite these findings, a clear theoretical understanding of the conditions that facilitate or discourage cross-cultural collaboration is seriously lacking.

Case Study Description and Historical Context

CLFN is an Anishinaabe (Ojibway) and Mushkegowuk (Cree) community located on the Kabinakagami River system, approximately 32 kilometres west of the primarily francophone town of Hearst, in northeastern Ontario (see Figure 4.1). At the time of this study, CLFN had a registered membership of 1,419 individuals, with approximately half of these members living on the reserve (Statistics Canada 2007). CLFN community incomes are derived from a mix of wage employment, transfer payments, and land-based harvesting activities, including seasonal hunting and fishing, with winter trapping still being an important activity (Hamilton et al. 2008). The CLFN traditional territory overlaps primarily with the OMNR District of Hearst, and the Hearst Sustainable Forest Licence (SFL), held by Hearst Forest Management Inc. (HFMI). HFMI is a shareholder company owned jointly by Lecours Lumber Co. and Tembec Industries Inc. CLFN ancestors (known as the English River Band) were signatories to Treaty No. 9 in 1906.[1] The current reserve location at Calstock was established in 1943, to offer local First Nation people access to employment opportunities in the emerging forest industry (Hamilton et al. 2008).

For a variety of reasons, CLFN people did not participate in the forestry boom that occurred in the town of Hearst during the last century. Hearst continues to be primarily dependent upon forestry. Three large wood-manufacturing facilities were in operation at the time of this study in the Hearst area. Tembec operated a sawmill in Hearst, Lecours operated a sawmill on the CLFN reserve, and Columbia Forest Products manufactured hardwood veneer and particleboard in Hearst. Approximately 41 percent of the local labour force was employed by these principal operators and a variety of supporting businesses (Corporation of the Town of Hearst 2003). Aside from some CLFN people employed at the Lecours lumber mill, CLFN members were not employed consistently in any of the mills or for harvesting work. This economic disparity, paired with adverse impacts from resource development and the loss of traditional harvest activities, created distrust and a troubled relationship between CLFN, industry, and governments.

The province has always maintained that a literal interpretation of Treaty No. 9 (and the other numbered treaties) established only that "Indians are entitled to hunt, trap and fish on non-reserve treaty lands except where these lands are taken up from time to time for settlement, mining, lumbering, trading or other purposes" (EAB 1994: 352, quoting Crystal 169–70 of Ex. 209).[2] The only legislative and policy requirements for First Nation involvement in forestry in 2003 were outlined in the Forest Management Planning Manual (FMPM) (OMNR 1996; OMNR 2004) and Condition 77 of the Class Environmental Assessment for Timber Management on Crown Lands in Ontario (EAB 1994), which required OMNR district managers to facilitate negotiations to ensure more equal participation by Indigenous peoples in the economic benefits of timber harvesting. Both of these processes engendered much criticism from and frustration in First Nation communities throughout northern Ontario (McKibbon 1999; NAFA 1995b; Smith 2007).

Negotiation Process Description

In January 1997, harvesting on a CLFN members' trapline, which was perceived by the trapper to infringe their treaty rights, spurred the CLFN members and leadership to block the only access road to the Lecours lumber mill.[3] CLFN members felt that the requirements of Condition 77 were not being met and that the impacts of harvesting on their Aboriginal and treaty rights were not being taken seriously, while OMNR and industry representatives considered that a blockade was premature as the parties had already scheduled an upcoming meeting. During the brief three-day blockade, CLFN brought forward a long list of forest-related grievances, including the conditions and implementation of Lecours' existing land lease, impacts of timber harvesting on traplines and the need for trapper compensation, issues of water quality, and the lack of First Nation employment in forestry operations in the region. The broad range of issues raised by the CLFN, together with the Timber EA Condition 77 requirements, led to the initiation of a formal mediated negotiation process. This process included the OMNR District Manager and several staff, and Indian and Northern Affairs Canada representatives from Thunder Bay.[4] Also at the table were executives from the main forest industries operating on the Hearst Forest Management Area (including Lecours, Tembec, and Columbia Forest Products), representatives from HFMI, the CLFN Chief, councillors and several community members, and an experienced mediator[5] selected by the participants and funded by the OMNR. The chief and council hired a Vancouver-based lawyer with Aboriginal and treaty rights experience to assist them.

The formal negotiation process consisted of approximately eight facilitated meetings held over six months. By June 1997, the mediator, with input from the CLFN lawyer, had crafted a draft coexistence agreement, which OMNR and industry participants were prepared to sign. The draft agreement contained commitments to support training, scholarships, employment, and business development opportunities but no actual numbers—for example, no harvest volume amount to support a community harvesting enterprise, and no dollar amounts for training, scholarships, or planning support. Early on in the negotiation process, an industry spokesperson had made a verbal commitment that some form of compensation would be provided to CLFN trappers for harvesting on traplines and that this compensation would apply retroactively to the date of the blockade. While this commitment was still acknowledged by forest industry participants, eventually these participants proposed that economic development for the whole community was more appropriate and an agreement implementation committee should work out the details of any direct financial compensation to the First Nation or individual trappers. The parties revisited the draft agreement several times, first in 2000 and then again in 2001. In the latter stages of the negotiation, CLFN participants explained that a government-to-government negotiation process to define a consultation approach was imperative and that specific quantification of the other parties' harvest allocation and financial commitments was required. First Nation participants were ambivalent toward the agreement, as it did not recognize their jurisdiction as a sovereign nation or establish a duly mandated rights-based government-to-government process to define a community consultation approach. The draft agreement was essentially shelved by the participants in 2002 and never signed.

Negotiation Process Outcomes

The CLFN negotiation process was initiated in order to produce a formal agreement among the parties, but almost five years following the initiation of a negotiation process, each group of participants described how they held little hope that a signed agreement was still possible, or sufficient. The CLFN had largely discounted the utility of the draft agreement to address trapline harvesting compensation or broader jurisdictional concerns or for securing commitments for forestry-related economic development. Despite their disappointment about the lack of a signed agreement, the OMNR and industry continued to focus on making progress on what they saw as shared goals arising from the negotiation process, namely improving communication for forest management planning and supporting CLFN capacity building and economic development opportunities. The First Nation participants eventually accepted that, given the limited scope of the negotiation, their community's interests (at least temporarily and under the current forest policy framework) were at least partially served by collaborative economic development opportunities. The First Nation continued to strive for forest management planning improvements locally, while concomitantly pursuing the recognition of their jurisdiction in other forums.

Concurrently with the formal negotiation, all three parties undertook several joint projects and they continued to be involved in ongoing collaborative activities. For example, the three industry partners and CLFN established a joint-venture forest harvesting company, called Mammamattawa Inc., in 2000 with allocation commitments of 97,000 cubic metres per year from the Hearst SFL. When Mammamattawa Inc. faced serious financial and management challenges in 2003, industry representatives and CLFN members on the board of directors were working collaboratively to restructure and improve the company. OMNR, industry, and CLFN members also collaborated to undertake negotiations with labour union representatives related to harvesting commitments on the Kenogami SFL that were allocated to Mammamattawa by OMNR conditions in the SFL but not accessible due to collective agreement restrictions. Industry partners and OMNR contributed to a small forestry education scholarship for CLFN members and supported joint economic development field trips with CLFN. HMFI developed a trapper communication protocol to guide forest management planning consultation activities. The OMNR District office also supported an Aboriginal liaison position and an on-reserve Native Values office and tried to improve First Nation participation in forest management planning by experimenting with various approaches over the years, including putting CLFN leaders or consultants on the planning team, paying First Nation planning team members per diems, and hosting on-reserve plan consultation open house bingos. The CLFN, OMNR, and Nord-Aski, a local non-profit economic development corporation, were also engaged in the development of an interpretive centre and cultural tourism facility adjacent to Nagagamisis Central Plateau Signature Site, called Eagle's Earth.[6] Finally, OMNR and CLFN collaborated for Northern Boreal Initiative (NBI) planning.

Research Approach

In 2003, individuals who had been involved in the CLFN negotiation process were interviewed. This involved face-to-face interviews with 17 (out of a possible 21) individuals, including 8 CLFN band members and employees, 6 forest industry employees, and 3 OMNR District employees and consultants. Semi-structured interviews were conducted using an interview guide, with questions developed to understand the structure of the negotiation process, the participants' expectations and satisfaction with the process, participants' perceptions of the outcomes, and the factors participants felt facilitated or discouraged collaboration. Interviews were recorded, transcribed, and analyzed using qualitative data analysis software QSR NVivo 2. The lead chapter author also resided in CLFN for several weeks, which enabled her to observe partner meetings and to review previous meeting minutes and agreement drafts. Data analysis[7] followed a two-stage inductive and iterative coding approach to identify the categories of outcomes identified by participants, the core outcomes consolidated from the coding themes and categories, and the main factors affecting outcomes.[8]

Factors Discouraging Cross-Cultural Collaboration

Different Understandings of the Problem

The parties in the CLFN negotiation process sat down to "negotiate" very different things. First Nation participants expected that their Aboriginal and treaty rights would frame the negotiation. First Nation participants described how throughout the negotiation process the sovereignty aspects of their definition of the problem were never understood, accepted as legitimate, or deemed feasible by the other negotiation participants: "they didn't recognize us as the owners, or part owners of the treaty, part-owners of the resources; they're still not budging in that area" (First Nation 05).

First Nation participants also expected the negotiation to allow them to change this status quo. For example, one participant described their vision of sustainability and the response from other participants: "We have to start looking at how the land can sustain development, not looking at everything in terms of jobs or the economy . . . instead of putting up arguments every time we want to say something, well the competition says we can't do it, world market says we can't do it . . . even if you are a company man, you still need to listen to our views . . . don't just put up barriers to it as soon as it comes on the table" (First Nation 01). First Nation participants also noted that while the OMNR considered that more or improved communication and integration of First Nation "values" into forest management planning would address the community's concerns, they felt that the entire framework for understanding, evaluating, and deciding on forest sustainability needed to change to address the impact of forestry on trapping as "a whole way of life" (First Nation 02). First Nation participants explained that no shared understanding of "sustainability" had developed among the parties, and the negotiation process in the end was akin to "talking more in circles" (First Nation 06).

Conversely, the OMNR and industry participants framed the main problem requiring attention as the lack of "integration" of the First Nation into forest management planning, forestry employment, and forest-related economic development. One OMNR participant recognized that local-level negotiations were lacking "Ontario's interpretation of treaty rights" (OMNR 25), but they nevertheless maintained that these issues were simply not within their "mandate." Industry participants admitted they were uncertain about their legal requirements with respect to harvesting on First Nation traplines upon entering into negotiations, and they framed the problem as the First Nation not benefiting from forest industry employment, training, and business partnerships. This framing of the problem in turn required that they try to negotiate a "win-win solution . . . to solve the First Nation social issues in the community and provide some peace and security of wood supply" (Industry 10). The resolution outlined in the draft agreement, which the OMNR and industry participants largely defined and advocated for, was to enable the *coexistence* of timber harvesting and the traditional activities of the CLFN on the Hearst Forest by placing responsibilities on all parties to work together to bring about the gradual *integration* of the First Nation into the forestry workforce and forest management planning process. OMNR and industry participants described that they had little ability, given existing policies and structures, to address many of the First Nation's issues. However, the coexistence agreement framework was seen to support the forest planning communication and the forestry-business integration problems they could influence. Eventually, the OMNR and industry participants' definition of problem and solutions dominated, which reflected and maintained their perceived rights and responsibilities, as well as their institutional and economic power.

Different participants described that their divergent understandings of the problem had limited collaboration outcomes. For example, according to First Nation participants the scope of the collaboration process was so constrained by the other parties' narrower problem definition that, even as improvements in the relationships and communication occurred, the resulting draft agreement was almost meaningless for the First Nation community. CLFN's desire for change regarding *how* forests were managed and *who* should make decisions about their traditional lands was never fulfilled. As a result, much underlying discontent remained. The scope of the negotiation was also limited to "forestry" as the other parties defined it. First Nation participants' realizations about the limits of the negotiation process in addressing their definition of the problem only came after considerable time spent negotiating with the other parties. Eventually realizations about the dominating narrow problem definition, and the limitations of the current legislative and policy framework, together caused some First Nation participants to embrace a forest-related economic development approach.

OMNR and industry participants also considered that the First Nation's definition of the problem had negatively influenced collaboration outcomes. For example, the First Nation's continual efforts to expand the scope of the negotiation into areas beyond what OMNR and industry participants perceived was their responsibility were seen as distracting all the parties from achieving any progress on what the OMNR and industry considered the solvable problems. OMNR and industry participants believed that the First Nation needed to "separate the issues" (Industry 08, 11; OMNR 09, 13) in order to play a more effective role in forest management planning and be considered a reliable business partner. In addition the First Nation participants' emphasis on "numbers in the agreement" was a significant source of frustration for the OMNR and industry participants. Participants' different understandings of the most appropriate scope and problems to address contributed to the failure to reach a signed agreement, to the parties' ongoing relationship difficulties, and to participants' disappointment with the negotiation process outcomes.

Relationship Development and Disappointment

Another barrier to collaboration was the failure of improved relationships and communication to contribute to meaningful changes in how the parties acknowledged and addressed each other's concerns. Both First Nation and non–First Nation participants identified improved relationships and communication as an important outcome of the CLFN process. For example, First Nation participants described the increased willingness of OMNR and industry to work with the First Nation, industry commitment to meet with trappers prior to harvesting on traplines, OMNR support for First Nation forest management planning involvement, and creation of an OMNR-funded Aboriginal liaison position and Native Values office on reserve, as evidence of improved relationships. Importantly, however, all participants considered that "improved" relationships and better communication were not suitable or sufficient outcomes of an otherwise costly, time-consuming, and for the First

Nation at least, politically risky process. Several First Nation participants noted that collaboration, in the form of business partnerships or community involvement in forest management planning consultation, without a clear recognition of their sovereignty, touched on the community's political and rights-based advocacy efforts and was therefore contentious within the First Nation.

First Nation participants explained that the relationships that had developed with OMNR staff and forest industry managers had not transformed how the other parties understood their underlying worldview of forests, sustainability, or their rights and responsibilities to the forest. Nor had the relationships been able to transform the other parties' perceptions of what they ought to do to address the First Nations' underlying political concerns and issues. Many of the identified positive outcomes of collaboration, such as improved sharing of information and improved CLFN participation in forest management planning, were overshadowed by the First Nation participants' disappointment. All First Nation participants supported the fundamental view that "they [the other participants] just don't want to recognize our Aboriginal title and our treaty rights . . . so how are we going to build a relationship if people don't want to recognize these" (First Nation 06).

OMNR and industry participants also felt that relationships were unsatisfactory, but for different reasons. The reluctance of the community to ratify the agreement after the other parties had invested their time and money, together with the First Nation's continued emphasis on rights-based claims (including government-to-government negotiations) and quantification of commitments, constituted a major disappointment for OMNR and industry participants, which negatively impacted their perception of the relationship. According to these participants, much of the good work resulting from the negotiation was unrecognized and the potential for finding solutions for long-term coexistence was lost. All participants noted that their relationships had improved following the negotiation to a point where they could at least work collaboratively on narrower planning and economic development goals, even if the broader issues related to rights-based claims, meaningful consultation, and revenue sharing remained.

Factors Facilitating Cross-Cultural Collaboration

Supportive Leaders

All participants believed that the presence of supportive individuals in industry, OMNR, and the First Nation community directly contributed to the initiation of a partnership approach and advance of a collaboration process. Participants noted that industry managers were "very community minded people . . . that is the flavour of the Hearst industry" (Industry 08). The working relationship between the OMNR District office in Hearst and forest industry was also described as stronger than elsewhere in the province and that "this spilled over into a better relationship with Native people" (OMNR 25). Industry and OMNR participants also considered that the CLFN leadership had contributed to the collaboration outcomes first by "building the community vision" (OMNR 09) and later by working in a collegial manner to improve economic development for the whole of the region. First Nation participants considered that an influential industry representative had contributed to the development of the coexistence approach and swayed some of the other forest industry leaders to support it (First Nation 02). Most participants stated that the Aboriginal liaison position, which was held by a CLFN member, served an important communication and interpretation function. Above all, supportive leaders from each group played a large role in initiating the negotiation process, communicating expectations and views to other parties, sustaining collaboration attempts despite challenges, and including lessons learned in their future approaches and organizational practices.

Recognition of Interdependence and Development of Shared Goals

The negotiation process and ongoing interaction between the parties contributed to a realization among the forest industry, OMNR, and CLFN that they were interdependent and had shared connections to the surrounding forest. All parties primarily framed this interdependence in economic sustainability terms. For example, participants emphasized that the culmination of their discussions was the recognition that they needed to work together for the benefit and prosperity of both the First Nation and non–First Nation communities. This recognition of interdependence is evident in the following statements:

> So if we're going to be a part of [industry] here, if we're going to be a stakeholder that actually works beside [industry], I mean we're not going to want to destroy everything because we know they have needs too. They have to live. We have to live. If we could come up with a good way of doing that and working together, I think that everyone will be happy with that (First Nation 02).
> We need to get them [First Nation members] involved. . . . We need to be sharing some of the benefits from harvesting and merchandising . . . We need to get them involved in our workforce. . . . So we need to be educating, we need to be building capacity. It's

> our future workforce. We need to have economic opportunities for them. (Industry 10)

OMNR and industry participants, who resided and worked in Hearst, had a vested interest in the future of "their" community and the forest industry as a whole. It was noted that the Lecours sawmill road blockade and subsequent negotiation "woke Hearst up" (First Nation 14) to the realization that both communities would lose out on economic opportunities if they did not work together. To this end, the "development of a sustainable economy" (OMNR 09) became a shared goal among the parties emerging from their recognition of interdependence. A First Nation participant noted that some industry and OMNR participants were also advocating for revenue-sharing approaches or a sharing of power given the parties' interdependence (First Nation 03).

The recognition of interdependence also influenced the OMNR and industry participants' perceptions of their own responsibilities. These parties recognized that they had a responsibility to encourage a First Nation leadership role in economic development initiatives, not only because these met Condition 77 requirements but also because they were in the best interest of both communities. To explain, government and industry came to believe that CLFN was owed some form of legitimate "compensation" and that CLFN involvement in economic development was necessary for regional sustainability and development. Finally, while industry participants never provided the community with direct compensation, as described previously, they still acknowledged that operating on traplines increased the chances that they were going to encounter resistance and that the First Nation legitimately needed to be compensated for such impositions.

Participants described several ways in which their recognition of interdependence and shared goals contributed to collaboration outcomes. For example, First Nation participants felt that the most tangible outcomes of the collaboration process arose from the other parties' recognizing the First Nation as a legitimate and valuable partner that could contribute to the region. Examples of this recognition included the increased allocation of fibre to the CLFN community; improvements in involving CLFN in forest management planning; industry's commitment to meet with trappers prior to initiating any harvesting on traplines; and OMNR's concerted efforts to generate economic development activities.

It was also clear that the parties' recognition of interdependence had influenced how shared goals were formulated and that this made them more acceptable to the First Nation participants. In particular, the goal of economic development was approached as a long-term, regional "partnership" directed by and responsive to the First Nation community. Pursuing a shared economic development goal was also likely more acceptable to the First Nation because the First Nation had maintained some power from choosing to not sign the draft agreement. The First Nation was still able to advocate politically for their rights-based claims in other forums or when required to do so with the OMNR and industry participants. At the same time, First Nations were also able to use their enhanced position to leverage support from their "partners" to work toward other more substantial policy changes and contribute to broader community development goals (e.g., industry support for resource revenue sharing). OMNR participants expressed the view that the willingness of community leaders to embrace mutually beneficial economic development opportunities—even for projects that were controversial or not accepted by other Indigenous communities[9]—indicated the progress that had been made in "separating the issues" in order to make headway on more narrowly defined but mutually acceptable goals.

Discussion

Examining the outcomes of cross-cultural natural resource management collaboration and the factors that facilitate and discourage collaborative outcomes has only just begun. This study, using an inductive approach, found that context-related variables (i.e., attributes beyond the control of participants) such as legislative and policy constraints, limitations placed on the scope of the collaboration by more powerful parties, the attitudes and leadership skills of participants, and perceived interdependence of participants, may be more important to cross-cultural collaboration outcomes than the internal characteristics of the collaborative process. These findings confirm that any attempt to evaluate cross-cultural collaboration should recognize that context, process, and outcomes are interrelated (Ansell and Gash 2007; Marshall et al. 2010; Plummer and Armitage 2007).

This study identifies two factors that limited cross-cultural collaboration, namely, the parties' different understandings of the problem and the limitations of relationship building. Both of these factors reflect several systemic institutional problems discouraging successful cross-cultural collaboration—for example, the lack of policy clarifying the obligations of the province regarding the Aboriginal and treaty rights of First Nation members, and disagreement between First Nation and non–First Nation participants about the validity and appropriateness of forest management legislation and planning. These results are similar to the findings of other authors, including Richards et al. (2004), Bidwell and Ryan (2006), and Marshall et al. (2010).

In this case, the parties' divergent understandings of the problem to be addressed significantly limited collaboration, but it would be overly simplistic to attribute process outcomes simply to the divergent understanding of the scope, because it masks the underlying worldview differences that explain why the parties did not define the scope the same way. CLFN experience demonstrates that where value and power differences exist and rights are contested, the less powerful parties' scope and problem definitions are constrained. Significant components of the First Nation participants' problem definition were excluded or reconstituted, and thus the worldviews underlying their definition of the problem were also not acknowledged.

It is also unlikely that the participants in this study would have been able to frame a mutually acceptable goal without engaging in substantive negotiations, and without experimenting with other less acceptable goals and, importantly, without the First Nation being able to explore the limits of the other parties' problem definition. Undertaking discussion and joint actions with a contested scope also allowed the First Nation to retain their power to present an alternative discourse based on their worldview, and it allowed all participants to recognize that the other parties' problem definitions were different. A shared understanding of the scope may not be a reasonable precondition for cross-cultural collaboration; however, some more limited version of a shared problem definition, or the gradual development of "win-win" or superordinate goals (Moore 1996), may be an *outcome* of the process.

This study suggests that it may be important for cross-cultural collaboration participants, and those seeking to evaluate cross-cultural collaboration, to focus on how divergent problem definitions and worldview differences are attended to and how mutual understanding develops. A willingness to work on shared goals seemed to be facilitated by the OMNR and industry participants accepting that the First Nation's trapping concerns were legitimate, and that the region and Hearst's growth and sustainability depended on involving the First Nation as a partner in economic development (even if the participants did not agree on how to do this).

The second factor limiting cross-cultural collaboration was the failure of improved relationships to transform how the groups acknowledged each other's main concerns. In particular, for the First Nation participants, developing a relationship should have resulted in policy changes for revenue sharing or meaningful consultation that would have included greater First Nation autonomous decision making. In the literature, it has been recognized that face-to-face dialogue and good faith negotiations contribute to trust building and improved relationships among stakeholders (Ansell and Gash 2007; Connick and Innes 2003; Wondolleck and Yaffee 2000), which leads to a shared understanding of the problem. Consequently, participants' perceptions of the relationship or other measures of improved stakeholder relations are often considered an "outcome" that can be used to evaluate collaboration (Cullen et al. 2010). Previous authors contend that building trust and developing relationships are outcomes of collaboration and preconditions for collaboration. For example, as communication and collaboration increases, trust increases, relationships are built, and more complex and genuine (equitable) forms of collaboration emerge (Wyatt et al. 2010).

In the collaborative process in this study, understanding and relationships increased as parties shared their goals, interests and perspectives, but the parties continued to have very different understandings of what was a valid goal and how to achieve it, and this negatively affected relationships and trust.

Our findings suggest that "improved relationships" are not a useful criterion for evaluating cross-cultural collaboration, and the presence of improved relationships may not reflect participants' true experiences of the equity of cross-cultural collaboration. It may be more useful to understand whether collaborative relationships served to facilitate achievement of the participants' diverse goals. For First Nation participants this may mean determining if improved relationships contributed to meaningful changes in how the other parties acknowledged and addressed their interests or concerns. Researchers interested in evaluating cross-cultural collaboration should endeavour to explore if the participants' relationships influenced the definition of the problem and how to resolve it, not just whether participants developed relationships (e.g., Cullen et al. 2010; Plummer and FitzGibbon 2006). A truly shared understanding of each other's perspectives in a cross-cultural context seems to require what other authors have called the *acknowledgement* of different worldviews or the "recognition that other disputants' views are valid and credible" (Elliott, Gray and Lewicki 2003: 428). This appears to be necessary for collaboration efforts to move beyond what participants in the current study described as "talking in circles."

The factors that facilitated cross-cultural collaboration outcomes in this study, including the presence of supportive leaders and the participants' ability to recognize their interdependence, are also consistent with the collaboration literature. The attitudes and skills of participants are identified among many other social and relational contextual factors that influence collaboration outcomes in several studies (see Table 4.2). The case of challenging cross-cultural collaboration in Hearst–Constance Lake forest management suggests that incentives are needed for government leaders to champion long-term collaboration. For example, resources and time could be allocated to information-sharing discussions with First Nation and forest industry leaders. Representatives from the CLFN, OMNR, and forest industry in Hearst have continued to meet quarterly, purely for the purposes of checking in, sharing information, and discussing issues or concerns. In order to bolster leadership capacity, district-level OMNR,[10] industry representatives and interested First Nation communities could also organize and support collaboration and conflict resolution training workshops with special emphasis on modules relevant to cross-cultural collaboration requirements. This would help increase participants' core competencies in this area.

The recognition of interdependence and development of shared goals in this case supports contentions in the literature that social capital and social learning are required for parties to be open to new ideas and to become adaptive in collaboration (Plummer and FitzGibbon 2006). However, whereas this literature finds that "deliberation that enables social learning may produce social capital, both of which are requisite for adaptive co-management" (Plummer and FitzGibbon 2006: 56), this study suggests that social capital and a shared commitment to place may already need to exist among the parties in cross-cultural collaboration, if more powerful parties are to expand their perceptions of their roles and responsibilities related to shared goals. Fostering a place-based identity, developing informal and formal social networks, and supporting intercultural liaison positions should be encouraged so that cross-cultural collaboration participants' deliberations are more likely to lead to the development of shared goals and shared perceptions of responsibilities.

The factors discouraging successful cross-cultural collaboration in this study also underscore an important conclusion in the literature that cross-cultural collaboration does not necessarily emerge because Indigenous peoples have a seat at the negotiation table (Ebbin 2011; Edmunds and Wollenberg 2001; Coombes and Hill 2005). However, this study suggests that cross-cultural collaboration processes can clarify the parties' scope, expectations, and understandings and contribute to the development of shared goals and collaborative projects at the local level.

As community representatives struggle with the instability in the forest industry, many are collaborating to advance alternative institutional arrangements to enable local communities to exercise more control over Crown forests (as illustrated by Palmer and Smith, Chapter 2; MacLellan and Duinker, Chapter 6; and Lachance, Chapter 5). The town of Hearst, the municipality of Mattice–Val-Côté, and the local forest industries, together with CLFN, worked together in 2010 to develop and promote an alternative local tenure model for the Hearst Forest, during the forest tenure reform process leading up to the creation of the Ontario Forest Tenure Modernization Act (Grech 2010). This model, which the participants characterize as a community forest, is intended for implementation as an Enhanced Sustainable Forest Licence, one of the two new tenure models possible as an outcome of tenure reform (see Harvey, Chapter 3, as well as Palmer and Smith, Chapter 2, for further details concerning tenure reform in Ontario and implications for northern groups). Both a CLFN and a Hearst representative were officially installed as ex officio non-voting board members of HFMI in 2010. It is clear that some institutional change, while it may still fall short of Indigenous peoples' desires, can emerge as an acceptable approach to local needs, as a result of sustained cross-cultural collaboration.

The CLFN collaboration process initiated in 1997, and the outcomes that were achieved, despite their limitations, provide a strong foundation for further collaboration among CLFN, OMNR, local municipalities, and the forest industry for the refinement and eventual implementation of a community forestry model. It remains to be seen if First Nation efforts to regain autonomy over their traditional territories are completely compatible or synergistic with arrangements developed under the ESFL model or if this new institution can address them. Undoubtedly, these efforts will require some form of collaboration, which, as this study finds, requires much greater attention paid to its implementation, including developing a better understanding of factors facilitating and limiting collaborative outcomes.

Notes

1 Treaties are agreements between the Crown (the government) and First Nations, in which the First Nations exchanged some of their interests in their traditional territories for various kinds of payments and promises from the Crown.

2 It has only been since Aboriginal and treaty rights were recognized and affirmed in the Constitution Act (1982) and with the decisions of *Haida Nation v. British Columbia (Minister of Forests) and Weyerhaeuser,* 2004 S.C.C. 73, and *Taku River Tlingit First Nation v. British Columbia (Project Assessment Director),* 2004 S.C.C. 74, that the province has acknowledged that they are legally obligated to respect treaty rights and have a duty to consult in good faith and possibly accommodate the concerns of an affected Indigenous group in keeping with the "honour of the Crown" (Ontario 2006: 514) when they contemplate a resource development activity which potentially infringes Aboriginal and treaty rights. The FMPM First Nation consultation requirements have remained fairly consistent since 1994 and were not altered after the province produced Draft Consultation Guidelines in 2006.

3 The Lecours family established the sawmill at Calstock, on reserve land leased from CLFN, in approximately 1960. At the time of this research, the Lecours sawmill was one of Ontario's largest and one of the last independently owned sawmills and held an allocation of 480,000 cubic metres (Hearst Public Library 2006). The lease agreement between Lecours and CLFN was not publicly available, but community members noted that the band leases the land to Lecours and a condition requires that at least 50 percent of the employees in the sawmill must be from CLFN. In 1995, a survey found that 65 CLFN members were employed at the Lecours sawmill but the total number of employees was not reported (NAFA 1995a).

4 Crown lands and natural resources, including forests, are the responsibility of the provincial government, specifically the OMNR, who manages them on behalf of all Ontarians. First Nation federally reserved lands (reserves) in Ontario fall within the core of federal responsibilities. Indian and Northern Affairs Canada representatives attended formal negotiation meetings, but they were not available to be interviewed for this study.

5 The mediator was Alan Grant, a well-respected Toronto-based lawyer and university professor with previous Aboriginal resource dispute resolution experience in the Magpie Agreement of 1989 with the Michipicoten First Nation.

6 Eagle's Earth planning was underway at the time of this study. The $12 million tourism facility, located approximately twenty minutes west of CLFN on the Shekak River, was officially opened in 2007. The facility was only operated by CLFN for a few years before it faced economic difficulties and closed.

7 Data collection and analysis procedures are described in detail in Casimirri 2013.

8 In the sections to follow, the interviewees' group affiliation and a number identifier follow direct quotes from participants in order to maintain confidentiality.

9 For example, CLFN was taking a lead role in the development of the Eagle's Earth cultural tourism facility, which was adjacent to an expanded protected area (Nagagamisis Central Plateau Signature Site), even though the protected area expansion had come out of a process that had previously been opposed by First Nation communities. Other Indigenous communities in the area were refusing to be involved in the project.

10 The OMNR divides the province into administrative regions, and then further into districts. Supervision of the Hearst SFL falls under the Hearst District OMNR office, located in Hearst.

References

Abelson, J. P-G. Forest, J. Eyles, P. Smith, E. Martin, and F-P. Gauvin. 2003. "Deliberations about Deliberative Methods: Issues in the Design and Evaluation of Public Participation Processes." *Social Science and Medicine* 57: 239–51.

Ansell, C., and A. Gash. 2007. "Collaborative Governance in Theory and Practice." *Journal of Public Administration Research and Theory* 18 (4): 543–71.

Armitage, D., F. Berkes, A. Dale, E. Kocho-Schellenberg, and E. Patton. 2011. "Co-Management and the Co-Production of Knowledge: Learning to Adapt in Canada's Arctic." *Global Environmental Change* 21: 995–1004.

Beckley, T., J. Parkins, and S. Sheppard. 2005. *Public Participation in Sustainable Forest Management: A Reference Guide.* Edmonton: Sustainable Forest Management Network.

Beierle, T., and D. Konisky. 2000. "Values, Conflict, and Trust in Participatory Environmental Planning." *Journal of Policy Analysis and Management* 19 (4): 587–602.

Bell, C. and D. Kahane. 2004. *Intercultural Dispute Resolution in Aboriginal Contexts.* Vancouver: University of British Columbia Press.

Bidwell, R.D., and C.M. Ryan. 2006. "Collaborative Partnership Design: The Implications of Organizational Affiliation for Watershed Partnerships." *Society and Natural Resources* 19: 827–43.

Booher, D.E., and J.E. Innes. 2002. "Network Power in Collaborative Planning." *Journal of Planning Education and Research* 21 (3): 221–36.

Carlsson, L., and F. Berkes. 2005. "Co-Management: Concepts and Methodological Implications." *Journal of Environmental Management* 75: 65–76.

Casimirri, G. 2013. "Outcomes and Prospects for Collaboration in Two Aboriginal and Non-Aboriginal Forest Management Negotiations in Ontario." Ph.D. diss., University of Toronto.

Castro, A.P., and E. Nielsen. 2001. "Indigenous People and Co-management: Implications for Conflict Management." *Environmental Science and Policy* 4 (4/5): 229–39.

Connick, S., and J.E. Innes. 2003. "Outcomes of Collaborative Water Policy Making: Applying Complexity Thinking to Evaluation." *Journal of Environmental Planning and Management* 46 (2): 177–97.

Coombes, B., and S. Hill. 2005. "'Na whenua, na Tuhoe. Ko D.o.C. te partner': Prospects for Comanagement of Te Urewera National Park." *Society and Natural Resources* 18: 135–52.

Corporation of the Town of Hearst. 2003. *Corporation of the Town of Hearst's Economic Development Strategic Plan: Perspective 2020 InSight.*

Cullen, D., G. McGee, T. Gunton, and J.C. Day. 2010. "Collaborative Planning in Complex Stakeholder Environments: An Evaluation of a Two-Tiered Collaborative Planning Model." *Society and Natural Resources* 23: 332–350.

Davis, N.A. 2008. "Evaluating Collaborative Fisheries Management Planning: A Canadian Case Study." *Marine Policy* 32: 867–76.

Docherty, J. 2004. "Narratives, Metaphors and Negotiation." *Marquette Law Review* 87 (4): 847–51.

Donoghue, E., S. Thomson, and J. Bliss. 2010. "Tribal-Federal Collaboration in Resource Management." *Journal of Ecological Anthropology* 14 (1): 22–38.

Ebbin, S.A. 2011. "The Problem with Problem Definition: Mapping the Discursive Terrain of Conservation in Two Pacific Salmon Management Regimes." *Society and Natural Resources* 24: 148–64.

Edmunds, D., and E. Wollenberg. 2001. "A Strategic Approach to Multistakeholder Negotiations." *Development and Change* 32: 231–53.

Elliott, M., B. Gray, and R. J. Lewicki. 2003. "Lessons Learned about the Framing and Reframing of Intractable Environmental Conflicts." In *Making Sense of Intractable Environmental Conflicts: Concepts and Cases*, eds. R.J. Lewicki, B. Gray, and M. Elliott, 409–435. Washington, DC.: Island Press.

[EAB] Environmental Assessment Board. 1994. *Reasons for Decision and Decision: Class Environmental Assessment by the Ministry of Natural Resources for Timber Management on Crown Lands in Ontario.* Toronto: Queen's Printer for Ontario. http://www.mnr.gov.on.ca/en/Business/Forests/2ColumnSubPage/STEL02_179249.html.

Gray, B. 1989. *Collaborating: Finding Common Ground for Multiparty Problems.* San Francisco: Jossey Bass.

Grech, R. 2010. "Hearst Area Communities Develop Tenure Model." *The Working Forest.* http://www.workingforest.com/hearst-area-communities-develop-tenure-model/.

Gunton, T., and J.C. Day. 2003. "The Theory and Practice of Collaborative Planning in Resource and Environmental Management." *Environments* 31 (2): 31–45.

Hamilton, S., L. Larcombe, J. Colson, P. Colson, P. Armitage, and D. Mackett. 2008. *Cultural Heritage Impact Assessment of the Fushimi Road Extension, north of Hearst, Ontario.* Report Prepared for the Constance Lake First Nation.

Hearst Public Library. 2006. "A Short History of the Hearst and Area Sawmills." http://www.scierieshearst.com/indexEn.html.

Innes, J.E. 2004. "Consensus Building: Clarification for the Critics." *Planning Theory* 3 (1): 5–20.

Leach, W.D., and N.W. Pelkey. 2001. "Making Watershed Partnerships Work: A Review of the Empirical Literature." *Journal of Water Resources Planning and Management* 27 (6): 379–85.

Margerum, R. 2011. *Beyond Consensus: Improving Collaborative Planning and Management.* Cambridge, MA: MIT Press.

Marshall, K., K.L. Blackstock, and J. Dunglinson. 2010. "A Contextual Framework for Understanding Good Practice in Integrated Catchment Management." *Journal of Environmental Planning and Management* 35 (1): 63–89.

McCool, S., and K. Guthrie. 2001. "Mapping the Dimensions of Successful Public Participation in Messy Natural Resources Management Situations." *Society and Natural Resources* 14: 309–23.

McKibbon, G. 1999. *Term and Condition 77: Origins, Implementation and Future Prospects.* http://mckibbonwakefield.com/pdf/12%20-%20Term%20and%20Condition%2077%20-%20Origins,%20Implementation%20and%20Future%20Prospects.pdf.

Moore, C.W. 1996. *The Mediation Process: Practical Strategies for Resolving Conflict*. San Francisco: Jossey-Bass.

Muro, M., and P. Jeffrey. 2008. "A Critical Review of the Theory and Application of Social Learning in Participatory Natural Resource Management Processes." *Journal of Environmental Planning and Management* 51 (3): 325–44.

Natcher, D., S. Davis, and C. Hickey. 2005. "Co-Management: Managing Relationships, Not Resources." *Human Organization* 64 (3): 240–50.

[NAFA] National Aboriginal Forestry Association. 1995a. *An Assessment of the Potential for Aboriginal Business Development in the Ontario Forest Sector*. Ottawa: National Aboriginal Forestry Association. http://www.nafaforestry.org/ontforestsector/.

———. 1995b. *Aboriginal Participation in Forest Management: Not Just Another Stakeholder*. Ottawa: National Aboriginal Forestry Association. http://nafaforestry.org/pdf/stakeholder.pdf.

NAFA/IOG. 2000. *Aboriginal – Forest Sector Partnerships: Lessons for Future Collaboration*: A Joint Study by the National Aboriginal Forestry Association and the Institute on Governance. Ottawa: National Aboriginal Forestry Association. http://www.nafaforestry.org/nafaiog/nafaiog3.php.

Ontario. 2006. *Draft Guidelines for Ministries on Consultation With Aboriginal Peoples Related to Aboriginal Rights and Treaty Rights*. https://www.ontario.ca/page/draft-guidelines-ministries-consultation-aboriginal-peoples-related-aboriginal-rights-and-treaty.

[OMNR] Ontario Ministry of Natural Resources. 1996. *Forest Management Planning Manual for Ontario's Crown Forests*. Toronto: Queen's Printer for Ontario.

———. 2004. *Forest Management Planning Manual for Ontario's Crown Forests*. Toronto: Queen's Printer for Ontario. http://www.mnr.gov.on.ca/en/Business/Forests/2ColumnSubPage/286582.html.

Passelac-Ross, M., and P. Smith. 2013. "Accommodation of Aboriginal Rights: The Need for an Aboriginal Forest Tenure." In *Aboriginal Peoples and Forest Lands in Canada*, edited by D.B. Tindall, R.L. Trosper, and P. Perreault, 129–50. Vancouver: University of British Columbia Press.

Pinkerton, E. 1989. "Attaining Better Fisheries Management through Co-Management Prospects, Problems and Propositions." In *Co-operative Management of Local Fisheries: New Direction in Improved Management and Community Development*, edited by E. Pinkerton, 3–33. Vancouver: University of British Columbia Press.

Plummer, R., and D. Armitage. 2007. "A Resilience-Based Framework for Evaluating Adaptive Co-management: Linking Ecology, Economics and Society in Complex World." *Ecological Economics* 61 (1): 62–74.

Plummer, R., and J. FitzGibbon. 2004. "Some Observations on the Terminology in Co-operative Environmental Management." *Journal of Environmental Management* 70 (1): 63–72.

———. 2006. "People Matter: The Importance of Social Capital in the Co-management of Natural Resources." *Natural Resources Forum* 30: 51–62.

Richards, C., K. Blackstock, and C. Carter. 2004. *Practical Approaches to Participation*. Socio-Economic Research Group Policy Brief No. 1. The Macaulay Institute. http://www.macaulay.ac.uk/ruralsustainability/SERG%20PB1%20final.pdf.

Robinson, E. 2010. "The Cross-Cultural Collaboration of the Community Forest." *Anthropologica* 52 (2): 345–56.

Selin, S., and D. Chavez. 1995. "Developing a Collaborative Model for Environmental Planning and Management." *Environmental Management* 19 (2): 189–95.

Shchepanek, M.J. 1971. "Trading Posts of the Moose-Michipocoten Trade Route." *Canadian Geographic Journal* 82 (2): 66–69.

Sheppard, S. 2005. "Participatory Decision Support for Sustainable Forest Management: A Framework for Planning with Local Communities at the Landscape Level in Canada." *Canadian Journal of Forest Resources* 35: 1515–26.

Shindler, B., K.A. Cheek, and G.H. Stankey. 1999. *Monitoring and Evaluating Citizen-Agency Interactions: A Framework Developed for Adaptive Management.* General Technical Report PNW-GTR-452. Portland, OR: U.S. Department of Agriculture, Forest Service, Pacific Northwest Research Station.

Smith, P. 2007. "Creating a New Stage for Sustainable Forest Management Through Co-Management with Aboriginal People in Ontario." Ph.D. diss., University of Toronto.

Statistics Canada. 2007. *Constance Lake 92, Ontario (Code 3556095)* (table). *2006 Community Profiles.* 2006 Census. Statistics Canada Catalogue no. 92-591-XWE. Ottawa. http://www12.statcan.ca/census-recensement/2006/dp-pd/prof/92-591/index.cfm?Lang=E.

Susskind, L., and J. Cruikshank. 1987. *Breaking the Impasse: Consensual Approaches to Resolving Public Disputes.* New York: Basic Books.

Tipa, G., and R. Welch. 2006. "Comanagement of Natural Resource: Issues of Definition From an Indigenous Community Perspective." *Journal of Applied Behavioral Science* 42 (3): 373–91.

Waage, S. 2001. "(Re) Claiming the Watershed: Property Lines, Treaty Rights, and Collaborative Natural Resource Management Planning in Rural Oregon." Ph.D. diss., University of California, Berkeley.

Wondolleck, J., and S. Yaffee. 2000. *Making Collaboration Work: Lessons from Innovations in Natural Resource Management.* Washington, DC: Island Press.

Wyatt, S. 2008. "First Nations, Forest Lands, and 'Aboriginal Forestry' in Canada: From Exclusion to Comanagement and Beyond." *Canadian Journal of Forest Research* 38: 171–80.

Wyatt, S., J.-F. Fortier, G. Greskiw, M. Hébert, D. Natcher, S. Nadeau, P. Smith, and R. Trosper. 2010. *Collaboration Between Aboriginal Peoples and the Canadian Forestry Industry: A Dynamic Relationship.* A State of Knowledge Report. Edmonton: Sustainable Forest Management Network.

Northeast Superior Regional Chiefs' Forum (NSRCF): A Community Forestry Framework Development Process

Colin Lachance

The Chiefs of the Northeast Superior Regional Chiefs' Forum (NSRCF) have been overseeing the advancement of a regional forestry agenda within the Northeast Superior portion of Ontario since 2006. Brunswick House First Nation, Chapleau Cree First Nation, Michipicoten First Nation, Missanabie Cree First Nation, and Hornepayne First Nation are founding members of the NSRCF. Chapleau Ojibwe First Nation is also located within the regional geography of Northeast Superior and has been kept apprised of NSRCF developments on an ongoing basis. The process has focused more on building relationships than on developing products, and the NSRCF has set several goals in this regard. The first is to build greater First Nation unity through a return to traditional Indigenous teachings that includes re-establishing kinships and building mutual respect and trust between the NSRCF member First Nations and adjacent municipalities through a comprehensive cross-cultural awareness process. As well, efforts are ongoing to strengthen partnerships with area forest companies such as Tembec (Chapleau sawmill) and Rentech (Wawa pellet plant) in support of developing a more robust First Nation economic agenda, using government policy commitments as leverage.[1] Last but not least, the NSRCF is committed to developing a better government-to-government relationship with Ontario as prescribed by the Supreme Court of Canada.[2] Central to all of this work is the need for area First Nations to become equal partners within Canada's constitutional fabric and to return to their traditional role as stewards of the land, consistent with Indigenous theology and prophecy.

A Principle-Based Approach

The NSRCF chose a principle-based approach as a means of facilitating consensus building among various process stakeholders and shareholders. This approach encourages addressing racial tensions created by history, misinformation, and fears associated with accommodating the advancement of the Aboriginal rights–based agenda (see Bopp and Bopp 2001). The guiding principles (i.e., justice, interconnectedness, unity, reconciliation, convergence, and ethics) were selected to facilitate transformative change through healing, empowerment, and collaboration. The NSRCF process recognized that court direction regarding reconciliation drives the need for transformative change (see Imai 2013). Increasingly prescriptive court decisions in critical areas such as the Crown's fiduciary obligation and duty-to-consult are reinforcing a growing understanding that the legal, political, and economic costs of ignoring the unfinished Aboriginal constitutional agenda outweigh the cost of solutions. Furthermore, these jurisprudence-driven opportunities are consistent with global societal trends that indicate a blending of mainstream values and traditional Indigenous values. Examples include movements toward greater environmental vigilance, economic-environmental-social triple bottom line planning, multicultural acceptance including spirituality in the workplace and other elements of balance from an Indigenous philosophical perspective. These attitudinal shifts are bringing people and communities together worldwide and are helping people focus on what they have in common rather than their differences (Naisbitt 1982). Some of these global trends that support community forestry are consistent with the spirit and intent of Canadian court rulings and are therefore important community advocacy and empowerment tools.

Tied to the reconciliation agenda is the concept of convergence, defined in this context as an approach that allows for the peaceful coexistence and merging of the underlying methodologies that support traditional Indigenous and western knowledge systems. Murphy, Chretien, and Morin in Chapter 11 also draw attention to the potential benefits of trying to clarify the points of contact that link settler/rural and Indigenous understandings and different ways of knowing in forest resource systems. Within the scientific domain, the growth of adaptive management as a planning tool is a key example given that it is built on the need for ongoing observation and correction of assumptions, a key traditional Indigenous concept. Within the cultural-spiritual domain, the advancement of spirituality from a personal rather than an institutional perspective and the rebirth of natural law as a foundation of this spirituality are important examples of convergence (Friesen 2012; Martin 2001), along with the renewal of the feminine spirit as women become increasingly involved in key leadership roles within various sectors of society.

NSRCF Strategic Objectives

Based on traditional Indigenous natural law teachings, and supported by an international best practices review of emerging resource stewardship trends undertaken by the NSRCF in 2011 (Lachance 2011a), a strategic agenda was developed that targets the need to find greater balance in the way in which natural resources within the region are managed and how the wealth from these resources are shared more equitably among various resource stakeholders and shareholders. Central to this ongoing discussion is the concept of community forestry and the need for First Nations to re-assume their traditional responsibilities as stewards of the land. Ecotrust Canada's Conservation Economy agenda was endorsed as a strategic cornerstone of such an approach, supported by a regional centre of excellence wherein each member community, First Nation and municipality alike, would build core capacity in a critical area and share this capacity with the other member communities on a reciprocal basis (see Lachance 2011b). The Conservation Economy agenda would support new economic opportunities in a number of areas such as value-added forestry, value-added tourism, and non-timber forest products. The centres of excellence would be coordinated with each other through a virtual network hub in support of balanced attention being given to regional economic, environmental, social, and cultural priorities without pulling critical capacity out of the respective communities in support of a centralized regional approach.[3] This concept is expected to continue to advance in collaboration with regional municipal partners through their active involvement in the process, facilitated by ongoing NSRCF engagement efforts. Meanwhile, the advancement of the Indigenous cultural renewal agenda is specifically entrusted to the NSRCF Elders Council.

The NSRCF also identified the need to foster an expansion of community forestry thinking within Ontario in support of building unity, consistency, and momentum. This was initiated on a number of fronts, including the NSRCF becoming signatories and major advocates of the Northern Ontario Sustainable Community Forest Partnership (NOSCP) Charter (a process that is described in detail in Chapter 2 of this book), and by reaching out to the Province of Ontario, Nishnabe Aski Nation (NAN) treaty organization, Ontario Forest Industry Association, the Ontario Professional Foresters Association, various Indigenous and municipal communities, and associations with community forestry interests and/or opportunities. Wahkohtowin Development GP Inc., a company recently formed to advance NSRCF strategic partnership opportunities, has been particularly successful in fostering an expansion of community forestry thinking in its inaugural year. For example, it has been formally retained by NAN to assist in the development of a government-to-government relationship with the Province of Ontario from a reconciliation perspective and has garnered preliminary interest from Ontario regarding the merits of developing a provincial forestry centre of excellence discussion paper.

Successes to Date

The NSRCF relationship-building agenda (see Lachance 2014) has resulted in a number of successes that are transforming the way in which First Nations, municipalities, forestry companies, and government work together in the Northeast Superior region. Notably, the NSRCF principled approach was used by the Ontario Ministry of Natural Resources and Forestry to modernize its forest tenure policy framework, resulting in a provincial commitment to support and resource an Enhanced Sustainable Forest Licence (ESFL) pilot project in the Northeast Superior region. Included in the proposed regional geography of this pilot project is the amalgamation of the Magpie and Martel forests and adjacent areas where merited, with special consideration being given to the ecological integrity of the 700,000-hectare Chapleau Crown Game Preserve (CCGP). This protected area was established by the province in 1925 to protect wildlife populations such as moose and beaver. It straddles several existing forest licences areas, including the Martel and Magpie forests. There are no special policy measures in place to protect wildlife in this area, and recent surveys have demonstrated that wildlife numbers in the preserve are very low. A CCGP moose recovery meeting held in 2011 concluded that a stand-alone CCGP-based forest management approach would likely contribute to better wildlife protection. It is from this challenge statement that the NSRCF began to advance a principled approach to forest reform as an opportunity, resulting in the launching of the Northeast Superior ESFL (NS-ESFL) development process.

The NSRCF is also the recipient of the first forestry-based resource revenue–sharing pilot project in Ontario, and will complete its initial two-year term in 2017. Although the initial funding formula was developed unilaterally by the Government of Ontario and fell far short of First Nation expectations, the NSRCF agreed to use the process as a stepping stone toward the development of a more equitable long-term arrangement. Meanwhile, separate business-to-business agreements have been signed between the NSRCF and a number of forestry companies. These agreements are now being handled by Wahkohtowin and are expected to expand over time with the possibility of profit-sharing mechanisms and as First Nation financial needs increase. The NSRCF equity-based economic strategy looks at community financial shortfalls and attempts to fill these gaps through a combination of core government funding supplemented by business-to-business, profit-sharing (industry-based) and resource revenue–sharing (government-based) agreements. A regional approach to capacity building is being advanced as a means of dealing with specific Indigenous employment challenges with the assistance of a number of academic institutions and the Superior East Employment and Training Development Task Force. As discussed by other authors in this volume (e.g., Bullock, Teitelbaum, and Lawler, Chapter 1; Palmer and Smith, Chapter 2; MacLellan and Duinker, Chapter 6), such collaborative approaches led by First Nations, academic, and government institutions have often provided much-needed support for the initiation of community forests across Canada. A number of academic, research, and strategic partnerships continue to evolve in support of the NSRCF centre of excellence agenda.

With respect to economic opportunities, the NSRCF continues to advance a number of strategic forestry initiatives, including a regional forest resource licence acquisition process. With respect to NTFPs, the NSRCF is currently in its third year of assessing priority opportunities that initially identified blueberry, birch syrup, and mushroom harvesting as areas of high potential. A regional field assessment of viable birch stands was undertaken in 2015, followed by a birch tree–tapping pilot project in 2016. Funding was recently secured to purchase birch syrup equipment in support of a 750-tap operation.

On the environmental front, a First Nations–led forest management planning pilot project is underway in the region, and an elders-led and youth-supported land guardianship program has been developed. The NSRCF Elders Council has also developed a comprehensive natural law–based cultural renewal work plan to support this agenda. Most recently, NSRCF and Tembec have agreed to advance a Forest Stewardship Council (FSC) certification pilot project on the Martel Forest that will be based on the concept of free, prior, and informed First Nation consent as prescribed by the United Nations.[4] Such a mechanism is expected to provide NSRCF member communities with a strong voice in forest management planning, thereby contributing to an increased ability to discharge their traditional environmental stewardship responsibilities. The NS-ESFL conceptual business model, currently in final draft stage, identifies the creation of a regional forestry investment fund that would provide core funding in perpetuity for the guardianship program.

Lessons Learned

Effective transformative change initiatives tend to be complicated given that they usually require numerous process stakeholders to reach and maintain consensus over an extended period of time within an arena of fear and distrust that is usually further complicated by pre-existing polarized relationships. The NSRCF development process has proven particularly daunting since it was initiated and facilitated by an ad hoc organization representing the most marginalized, under-resourced, under-appreciated, and misunderstood of the regional partners. A principle-based approach has proven to be effective in guiding the various process stakeholders toward a common understanding.

Focused attention to relationship building within the NSRCF developmental process has led to a number of insights regarding the importance of emotional intelligence in resource decision making. The more intimidating transformative change elements of the process were much easier to advance when process stakeholders were given an opportunity to speak from the heart, particularly when discussing the common need to protect regional resources in the long term for the sake of future generations. This priority was facilitated by paying close attention to the wisdom of the elders as well as trends that harness the renewal of the feminine spirit. Such an approach was particularly useful in building compassion and consensus around the need to close the Indigenous social and economic gap, starting with marketing a more accurate understanding of the way in which First Nations have been treated historically under Canada's constitutional framework.

Another important lesson learned pertains to the way in which brinkmanship methods used by the NSRCF provided an invaluable empowerment catalyst for regional leaders and citizens alike. Not enough brinkmanship would result in continued complacency, while too much brinkmanship ran the risk of alienating those who need to become engaged as meaningful partners in support of a true community forest approach. Continuously finding a balance point as the process unfolded proved elusive at times given that the target audience continued to shift as the process advanced. There also came a time when the process reached a critical mass, whereupon the brinkmanship agenda moved from leading the process to supporting the process. This resulted in the creation of the Wahkohtowin corporation. Consistent with the Indigenous Medicine Wheel model of community development (see Bopp and

Bopp 2001), related NSRCF initiatives and pilot projects include community forestry governance (NS-ESFL conceptual business case), fiscal autonomy (resource revenue sharing and resource benefit sharing), economic development (business-to-business), government-to-government relationship building (free, prior, and informed First Nation consent), environmental stewardship (guardianship program), social development (employment gap analysis and job training strategy), cultural renewal (a return to ceremony, cultural asset mapping, and supporting information management), and reconciliation (on-going cross-cultural training). Many of these initiatives are poised to evolve into regional centres of excellence, to be established at the First Nation and hopefully at the municipal level.

Next Steps

It is hoped that as the aforementioned pilot projects advance and expand, they will become integrated into a comprehensive regional model, supported by a regional planning council with coordinated political oversight provided by the regional leadership, coordinated strategic oversight provided by regional community administrators and technical support provided by community economic development and other officers. This concept was originally proposed by the NSRCF in a discussion paper prepared several years ago (see Lachance 2011c) that identified the need to replicate the successes of these community-based forestry activities in other resource sectors, including the energy sector, since it is a major cost driver in the forestry sector, in support of developing a comprehensive, inclusive, participatory, and fiscally independent regional mechanism that serves the collective economic, environmental, social, and cultural interests of its citizens. In support of this end goal, Wahkowtowin recently accepted the mandate to advance the NSRC regional energy strategy and regional mining strategy (currently in their second year of advancement), along with supporting funding provided by key provincial agencies.

Notes

1 Notably, the Ontario Ministry of Natural Resources committed in 1994, and refreshed this commitment in 2003, to negotiate economic arrangements between First Nations and forest licence holders as one of a number of conditions required for the approval of the class environmental assessment of forest management on Crown land by the Ontario Environmental Assessment Board (see Declaration Order MNR-71 [2003]).

2 In his recent book, *Resource Rulers*, Bill Gallagher explains in detail how and why First Nations "have racked up the most impressive legal winning streak in Canadian history with well over 150 wins" (Gallagher 2012: back cover).

3 The Institute for Northern Ontario Research and Development (INORD), located at Laurentian University and specializing in social and economic research in northern Ontario, has signed a reciprocity agreement with the NSRCF that establishes it as the main academic partner in support of creating a centre of excellence network hub.

4 The free, prior, and informed consent agenda operationalizes the Crown's duty to consult and accommodate to avoid the infringement of Aboriginal inherent and treaty rights. The Boreal Leadership Council produced a document in September 2011 entitled *Free, Prior, and Informed Consent in Canada: A Summary of Key Issues, Lessons, and Case Studies Towards Practical Guidance for Developers and Aboriginal Communities,* which provides a solid overview of related challenges and opportunities.

References

Bopp, M., and J. Bopp. 2001. *Recreating the World: A Practical Guide to Building Sustainable Communities.* Calgary: Four Worlds Press.

Boreal Leadership Council. 2012. *Free, Prior, and Informed Consent in Canada: A Summary of Key Issues, Lessons, and Case Studies towards Practical Guidance for Developers and Aboriginal Communities*. Ottawa: Boreal Leadership Council. Accessed 17 June 2015. http://borealcouncil.ca/wp-content/uploads/2013/09/FPICReport-English-web.pdf.

Friesen, J.W. 2000. *Aboriginal Spirituality and Biblical Theology: Closer Than You Think.* Calgary: Detselig, 2000.

Gallagher, B. 2012. *Resource Rulers: Fortune and Folly on Canada's Road to Resources.* Waterloo: Bill Gallagher.

Imai, S. 2013. *The 2013 Annotated Indian Act: Indian Act and Constitutional Provisions.* Scarborough: Carswell.

Lachance, C. 2011a. *Natural Law in a Modern Context: A Comparative Assessment of Understanding and Opportunities.* Chapleau, Ontario, Canada.

———. 2011b. *Towards a Traditional Aboriginal Approach to Centre of Excellence Development.* Chapleau, Ontario, Canada.

———. 2011c. *Northeast Superior Forest Community Renewal Strategy Discussion Paper.* Chapleau, Ontario, Canada.

———. 2013. *NSRCF Vision Statement and Five Year Action Plan Results.* Chapleau, Ontario, Canada.

———. 2014. *Building a Government-to-Government Relationship between the NSRCF and the Government of Ontario: A Discussion Paper and 2014–2015 Action Plan.* Chapleau, Ontario, Canada.

Martin, A. 2001. *Journey into the Light*. Penryn, CA: Personal Transformation Press.

Naisbitt, J. 1982. *Megatrends: Ten New Directions Transforming Our Lives.* New York: Warner Books.

The Local Trap and Community Forest Policy in Nova Scotia: Pitfalls and Promise

Kris MacLellan and Peter Duinker

Nova Scotia stands at the cusp of the largest change to forest tenure in living memory. Large, industrial forest tenure licences have long been the principal means to harvest timber on Crown land. In 2011, the release of a formal natural resources strategy for Nova Scotia set in motion changes that have yet to be fully realized but that stand to revolutionize both the way in which citizens relate to their forests and the relationship of government with those forests. The strategy document, entitled *The Path We Share* (Government of Nova Scotia 2011), contained a single line of text that set forest stakeholders in Nova Scotia alight. On page 38 was a commitment for action on the part of the government to "explore ways to establish and operate working community forests on Crown land." Where had this extraordinary commitment come from? Who within the government had suggested it? Had that person thought through the ramifications of such a suggestion? These questions, among many others, galvanized those with interests in the forest.

A phase of intensive scrutiny followed concerning the notion of community forests. Among the actions commissioned by the Nova Scotia Department of Natural Resources (NSDNR) to explore the community forest concept was a project undertaken by Dalhousie University and the Nova Forest Alliance (NFA) in the summer of 2012. Its purpose was to bring together key stakeholders and analyze the community forest concept. This led interested parties to organize community forest advocacy groups to pressure the provincial government for official implementation of the idea. Consequently, the Government of Nova Scotia then entered into confidential talks with major forest landowners, onto which community forest advocates are projecting their hopes and expectations. In this chapter we explore the prospects for community forests in Nova Scotia and bridge relevant literature and research dimensions with operational realities. We also explore logistical and theoretical pitfalls that face the fledgling Nova Scotia community forest program and how these may be avoided during a formative period of policy reform.

Context

The current state of Nova Scotia's forests, from both biophysical and sociopolitical standpoints, is the result of decades of inadequate planning (Sandberg and Clancy 1996). The pulp and paper industry has been the major player in the Nova Scotia forestry sector for decades. The practice of managing forest resources for quantity over quality has left a legacy of landscape degradation. A century of "take the best, leave the rest" (a timber-harvest practice known as *highgrading*) and wholesale clearcutting for pulp and paper production have left much of Nova Scotia's Acadian forest in a severely degraded state (Salonius 2007). The effect of whole-tree harvesting in short rotation has been a deterioration of the resource base by most quantitative measures, including height, girth, species composition, increase of fire barrens, nutrient depletion, and shortages of merchantable timber (Goldsmith 1980). Environmental action groups frequently vilified the practice of clearcutting and aerial herbicide spraying as contributors to the perceived ruin (Chappell and Simpson 2010). Indeed, when Nova Scotians elected a government for the first time formed by the centre-left New Democratic Party, stakeholders in the province hoped that transformational change in forest management was forthcoming. When *The Path We Share* was released in August 2011, the inclusion of community forests was widely regarded as a promising step toward fulfilling the implicit promise of sustainability in the new government's mandate.

The Dalhousie-NFA study produced two documents, *Community Forests: A Discussion Paper for Nova Scotians* (MacLellan and Duinker 2012a) and *Advancing the Conversation on Community Forests in Nova Scotia* (MacLellan and Duinker 2012b), which were released around a forum event organized by us to inform the DNR project. The event, held at Dalhousie in June 2012, brought together a diverse group of forest stakeholders to raise awareness about the potential for community forests and to explore the shape they might take. Specialists from across Canada were brought in to address the forum, including notable representatives from environmental advocacy groups, academic institutions, the pulp and paper industry, forest research groups, municipal governments, and woodlot owners. While their interests were diverse, all participants expressed hope in seeing the community forest concept become successful and all pledged to support it.

A move toward community forests in Nova Scotia is evidently a change welcomed by many forest stakeholders. But why did the government decide that local control of forests was attractive? What influenced this decision? Stagnation in any industry, be it healthy or not, is undesirable. To call today's forest products industry in Nova Scotia healthy would be disingenuous, and there are several avenues the province might have taken in efforts to rejuvenate this struggling sector. One such direction was actually pursued: the construction of a large biomass-using electricity plant at Point Tupper in Cape Breton. While this does determine the final destination for forest yield, it does little to address issues of sustainability in the harvest itself. Nor is it an impetus to develop alternative non-timber forest products, reduce clearcutting, or help to regenerate the province's degraded forest condition—all of which are often associated with community forest principles and practice (see Bullock, Teitelbaum, and Lawler, Chapter 2). Properly implemented, a fully functioning community forest program could help to address all of these issues.

Pitfalls and Promise

For some, Nova Scotia may seem an odd case for a community forest network. In very rough terms, about 50 percent of the forest land is owned in small parcels, or woodlots, by individuals and small family companies. About 20 percent is owned in extensive holdings by large firms. The remainder—about 30 percent—is Crown land (MacLellan and Duinker 2012a). Compare this to British Columbia, where private ownership is relatively miniscule, and another hurdle is evident. Any movement forward will need to capture the imaginations of the public and the acceptance of Nova Scotia's influential private woodlot owners. Yet throughout the world, where community forests have been successfully used to repair degraded forest ecosystems and lift economically depressed areas out of poverty (Oh 1986), it is precisely these "odd cases" where community forests have their greatest positive effect. In South Korea, they were used to set the stage for economic stability and development after years of conflict stripped the country of its forests (Oh 1986). In India, community forests provided fuelwood to a population desperate for the basic necessities of life (Pagdee et al. 2006). In the United Kingdom, they are at the forefront of returning that country's forests to good health through an afforestation program after the island was nearly denuded (England's Community Forests

2005). Nova Scotia may have factors that complicate the path to community forests, but these are not insurmountable. If anything, the challenge could energize a movement forward.

The community forest movement cannot happen without collaboration. Small woodlot owners in Atlantic Canada have some limited experience in co-operative action (Teitelbaum et al. 2006). They have formed provincial federations, marketing boards, group ventures, and co-operatives to represent their concerns, negotiate market access, and encourage forest stewardship. These are all skills that will be invaluable to the future of community forests in this province (Sandberg and Clancy 1996; Clancy 2001; Teitelbaum et al. 2006). While Nova Scotia in particular is not a notable example of these collaborative forces achieving success, the high level of private ownership should not be seen as a hindrance but rather as a support. The challenge of incorporating the realities of Nova Scotia's high level of private land ownership into a provincial community forest model could result in a unique framework.

Size and Scale

The idea of community forests is spreading in Nova Scotia, and acceptance of the idea comes from multiple angles. Changes to the 1989 Nova Scotia Forests Act may be necessary, though this has not been debated publicly. There will be much wrangling about how to see the project come to fruition. Recently acquired by the province, the former Bowater lands alone constitute over 220,000 hectares of forested land suitable for community forests (Conrad 2012). This suggested community forest project is also ringed by communities that once depended economically on the pulp and paper industry. To form the new land purchase entirely into a single community forest pilot project would probably make for an inappropriately large area for a single community forest (given the spatial context). In British Columbia, over 95 million hectares in size as compared to Nova Scotia's 5.5 million hectares, and where community forests are prominent, the average licence is much smaller than what is available on the Bowater lands in totality—more than 100,000 hectares (BCCFA 2015; MacLellan and Duinker 2012b). As of mid-2014 in British Columbia, there are 57 actual and pending community forests (see Gunter and Mulkey, Chapter 8). Smaller community forests, such as North Island Community Forest and the Cherry Ridge Community Forest, are 2,392 and 1,081 hectares respectively (BCCFA 2015), while the two largest, in Fort St. James and Toba Inlet, are more than 100,000 hectares each (BCCFA 2015). This network of community forests contributes to approximately 2 percent of the provincial annual allowable harvest (BCCFA 2015). In Nova Scotia, a much smaller province with a variety of existing industrial tenures on Crown land, experimental community forests will doubtless have to begin with relatively small land bases.

Indeed, establishing large community forests in Nova Scotia perhaps contravenes some community forest management principles. The first two principles, local control and community benefit, may be compromised in a community forest greater in size than, say, 100,000 hectares. Moderately sized community forests, say in the range of 10,000 to 50,000 hectares, seem more appropriate for the Nova Scotian context. This range of sizes will allow for greater diversification and is more in keeping with the recommendations of Duinker and MacLellan (2012) calling for variety and experimentation (see below). Once initial pilots show success, the community forest program could then be spread to the rest of the province.

Falling into the Local Trap

In implementing the community forest idea, the Government of Nova Scotia may have seen a rising trend elsewhere toward devolution and local management and taken this opportunity to promote such change in the province's forests. To be sure, the four guiding principles of community forests established by the UN Food and Agricultural Organization (UNFAO 2012) (i.e., participation and local decision making, community benefit, multiple-use management, and ecological sustainability) are all concepts that fall within the ideological purview of a centre-left political organization, in this case the Nova Scotia New Democratic Party. It could be that the government of the day gravitated naturally toward a management structure in keeping with its philosophical leanings (see Harvey, Chapter 3). To suggest that the motivation for community forests comes from a purely political origin diminishes the practical and logical benefits that can be achieved through community-based natural resource management. Assuming that the economic, social, and environmental benefits of community forests are understood, it then behooves forest stakeholders to dig deeply into this idea to avoid making the mistakes of the past.

The potential for concern lies in determining the inherent value (if any) which may come from the movement toward localization of management in the Nova Scotia context. The presumption that community forest management is inherently superior to a larger and centrally organized system must surely be supported by evidence if it is to be used to advance systemic change. Otherwise, it will be tempting simply to state the obvious: decades of provincial governments and interchanging corporate stakeholders have been poor stewards of Nova Scotia's forests. Economically speaking, the industry is languishing. This is evidenced by the rash of mill closures or near-closures and the extraordinary steps and costs assumed by government to keep some mills operating (Leger 2011). Knowing this, another argument could be made that the government is simply suffering "closure fatigue" from the constant political and financial demands created by buttressing the industry's economic viability. This perspective envisions the community forest model as a way to solidify at least one dimension of the sector, perhaps hoping it will serve as a basis for a wider stabilization of the forest products industry in the province.

If the movement toward community forests is based solely on the assumption that local stakeholders will be inherently superior stewards of their environment, the Government of Nova Scotia runs the risk of falling into a "local trap." Such could be the fate of any public policy built on the assumption that the underlying truth of the matter is self-evident, rather than taking the time to unpack the obvious more fully. The problem is as subtle as it is complex. After more than a century of ubiquitous corporate, provincial, and federal control and related forest degradation, how could a move toward the local *possibly* be any worse? A small rural group democratizing the forest sector is easily presented in an overly romantic light, and the logic is seemingly self-evident: how could anyone intelligent enough to harvest a resource be so short-sighted as to deplete that resource at their own expense, against their own self-interest?

These questions are themselves contributors to "the local trap" (Brown and Purcell 2005). The notion that managing natural resources at the local scale is inherently superior to other scales—the essence of the local trap—found prominence in the Rio Earth Summit in 1992. Principle 22 of the Rio Declaration claimed that "local communities have a vital role in environmental management and development because of their knowledge and traditional practices" (United Nations Department of Economic and Social Affairs 1992). Canadian scholars have noted this challenge among community forest advocates, including practitioners, community members, and academics (Teitelbaum et al. 2006; Bullock and Hanna 2007). Such misguided assumptions overlook the reality that some communities may not want or be able to manage local forests (Duinker et al. 1994). The actions of the UN set in motion the movement toward recognition of the inherent goodness of local management, though it might never have gained traction if not for the onset of globalization. The dramatic shift toward global thinking that accompanied the rapid improvement of telecommunication technologies, international trade, and the internet paradoxically underscored the apparent intrinsic value of the local scale. As the term "global village" passed in and out of vogue, political ecologists coined the phrase "glocalization" to describe the new state of flux between global and local scales (Brown and Purcell 2005).

It was not until a decade had passed after Rio that development scholars Brown and Purcell (2005) named "the local trap" and began to identify the pitfalls that stemmed from wholeheartedly embracing community-scale management of natural resources. The "trap" was seducing development scholars and practitioners around the world with the promise of democratization and social justice, but the unexpected side effects manifesting themselves were in some cases demonstrably harmful. Brown and Purcell (2005) noted that the local was too often equated with "the good," but what "good" actually meant varied so wildly from context to context as to rob the term of its usefulness. Brown and Purcell (2005) suggested that the local trap captures development practitioners not only in their overt vision for local community development but also in their unconscious assumptions—where it can be most harmful.

Spreading caution about the dangers of the local trap should not be misinterpreted as a blanket warning against community-oriented management strategies—in Nova Scotia or elsewhere. In fact, being forewarned about the common problems involved in the devolution of power can be used to help Nova Scotia's newborn community forest program thrive and avoid common problems like environmental degradation and increased inequality. Aside from the above-mentioned pitfall (blind acceptance of the righteousness of localization), scholars associate two other common problems caused by the local trap. Mismatching of ends and means represents the second typical problem—the tendency to assume that a goal (greater democratization, egalitarianism, sustainability) is accomplished simply by embracing a particular strategy (localization). Nova Scotian planners, therefore, could "become sidetracked pursuing localization and become distracted from pursuing their real goal, whatever that might be. At the very least, this dynamic [could] cause planners to lose sight of their goal. In the worst case, it will subvert their goal, as when a planner who desires greater food democracy pursues localization that results in more oligarchical decision making" (Born and Purcell 2006: 196).

The final and perhaps most delicate hazard in the local trap is simply the failure to consider alternative scalar arrangements that may be more appropriate (Born and Purcell 2006). A Nova Scotian community forest strategy could, for example, be remiss in pursuing locality for locality's sake if there are provincial, regional, or national organizations that may be able to provide superior support or advice at key stages.

Nowhere were the consequences of the local trap more poignantly demonstrated than in Brazil during the 1980s. Ironically, the nation that would host the most influential environmental summit in history is home to one of the best examples of local forest governance gone wrong (Lutzenberger 1985). During the early 1980s, Brazil was nearing the end of a decades-long dictatorship. That regime, bent on returning control over Brazil's forest industry to the people, got caught up in romanticizing the peasant lifestyle and in a strong desire to bring the central Amazonian region into the economic fabric of the nation state (Brown and Purcell 2005). Agriculture was seen as the driver for economic growth, and the means to attain said growth was deemed to be local control over land and resources. The building of highways and the pressure to resettle coastal populations inland were key components of POLONOROESTE, an infamous project funded by the World Bank (Lutzenberger 1985). Government assistance for settlement and development of the forest industry led to a dramatic spike in deforestation in the Amazonian state of Rondonia. In 1980, only 3.1 percent of Rondonia's area had been deforested—by 1985 that figure reached 11.4 percent (Fearnside 1989). Government ineptitude and weak local governance structures also allowed failed agricultural projects simply to move farther inland, leaving deforested areas to lie fallow. By the time the government officially accepted the scope of the failure, a trend of deforestation, conflict, and environmental ruin had begun. This drift eventually saw an area of critical rainforest covering more than 600,000 square kilometres deforested (Malhi et al. 2008).

The local trap can cause real problems. Accepting that all scales are relative, and that local scalar arrangements need not be any more inherently bad than they are inherently good, progress is possible. In Brazil, investments in community forests are part of billions of dollars in government spending meant to support local community-oriented organizations (Nepstad et al. 2009). This funding is intended to manage natural resources in a way that promotes sustainability and halts the trend of deforestation (Nepstad et al. 2009). Community forests and local projects are now a part of the solution, fostered by a "glocal" worldview that accepts the value of management that spans spatial levels, frameworks, and jurisdictions. This kind of progress can work in Nova Scotia, as described in the following section.

Next Steps

As refurbishment of the forest tenure system in Nova Scotia progresses, the government has sought assistance from multiple sources. Outreach to academia and civil society has coincided with a groundswell of grassroots support, both for the implementation of community forests and the public repurchase of privately held tracts of forest land. Our consultation and research activities at Dalhousie University led to a document entitled *In Support of Community Forests: Recommendations to the Government of Nova Scotia* (Duinker and MacLellan 2012). In it, a vision is proposed in which the government would support a network of diverse community forests across the province. The vision bridges the realities of working in a province with a multitude of apparent constraints—those of geography, history, and economy—with an optimistic operational reality. Nova Scotians expressed their interest in community forests and identified the role of government as one of paving the way, setting the ground rules and institutional framework to facilitate the program and enabling a culture of collaboration without becoming overbearing or intrusive.

While a spirit of variety and experimentation is considered vital, a central set of principles recognizable in community forests all over the world guides community forest program development. These standards include community control, local benefit, multiple-use management, and sustainability (Duinker et al. 1994; Teitelbaum et al. 2006; Teitelbaum 2014). Part of what makes the potential of a community forest model so attractive is the idea's malleability. While the aforementioned principles remain, community forest projects would be as different from each other as are the communities that host them. As is done in British Columbia (a national leader in community forest management), community forests in Nova Scotia would be managed by the people who stand to benefit from them. Those individuals could be expected to direct benefits into local businesses and economies, and would likely support a stringent monitoring regime working to ensure that practice reflects principle. The report contained five recommendations upon which a vigorous program of community forest development could be based (Duinker and MacLellan 2012):

1. **Proceed now**. While unnecessary haste is certainly no positive quality, further procrastination in addressing the problems plaguing provincial forests is unacceptable. Development of the community forest program must begin immediately, with production of a framework for a government assistance program being a high priority.

2. **Promote diversity and rigorous experimentation**. A new, community-oriented model for forest management in Nova Scotia is as yet untested. A single model replicated across diverse jurisdictions could be a recipe for wider failure. One hundred years of experience with community forests across Canada indicates that context varies and each model needs to account for local conditions and needs. The opportunity to learn through adaptive management is not one to be missed.
3. **Engage broadly**. The recent steps taken by the provincial government (e.g., the 2012 community forest forum) should be seen as only the first step toward a larger consensus-building enterprise. A vigorous program of stakeholder engagement will create the greatest endurance of strategic vision.
4. **Establish support mechanisms**. Enthusiasm for community forests at the highest levels must be evident, and risk taking among community forest managers should be encouraged. Failure in this experimental process must not result in the death of an individual community forest, nor the larger program.
5. **Evaluate continuously**. The community forest projects under consideration now are no small undertakings. Review mechanisms, administered by a government department, should examine the successes and failures of the network as it evolves. As was brought to light during the Dalhousie forum event, this process should also celebrate Nova Scotia's newly democratized forest sector and capture the narrative of that success in a meaningful way.

These recommendations, delivered to then Minister of Natural Resources Charlie Parker in September 2012, form a strong basis for a new phase of forest governance in Nova Scotia. Since the end of direct involvement by Dalhousie and the Nova Forest Alliance, stakeholder groups energized by the forum event have mobilized to press the government to take action. Furthermore, embracing these recommendations has alleviated some potential hand-wringing around the local trap. A rigorous regime of experimentation, evaluation, and engagement could tip the scales against potentially ruinous over-localization.

Nova Scotia's Minister of Natural Resources uttered the term "community forest" in broadly consumed media for the first time in September 2012 (CBC 2012). This support reflects the minister's own remarks at the Dalhousie forum event, at which he proclaimed that "the status quo [in NS forestry] is not an option. And the collaborative approach of community forests is an attractive one" (MacLellan and Duinker 2012b: 5). Then Premier Darrell Dexter was quoted as stating in the legislature that Nova Scotians are "sick and tired of watching the future of the province being sold off to foreign interests" (NS 2012: 3618). This was in reference to the revelation (at the time) that the government had indeed become the primary bidding interest in purchasing the Bowater lands with the stated intent of ensuring they remained managed with local interests in mind (CBC 2012). Support at this level hinted at continued interest in community forests at some of the highest levels of power.

The labours of community groups, professional associations, and environmental NGOs, in addition to behind-the-scenes negotiations between government and industry, led to an announcement in December 2012 indicating the government's decision to pursue community forests in Nova Scotia. Over the course of two days, the province disclosed its formal intent of creating community forests, its decision to purchase the Bowater lands, and its goal of transforming a former Bowater mill into a centre for excellence in forest-related research (NSDNR 2012).

Creating a Path to Share

The shared path toward community forests is neither straight nor solitary. There are multiple ways to arrive at a final destination which may provide all parties to the process with greater satisfaction. This chapter has tried to provide insight into but a few potholes that may dot the path itself. The obstacles facing the initiative's fruition may be daunting, but they need not seem so overwhelming as to elicit a fear response—particularly if that response results in timidity, or half-measures, during the implementation years. We called the inclusion of community forests in *The Path We Share* a "special gift" (MacLellan and Duinker 2012a: 16.) to the citizens and communities of Nova Scotia. This sentiment was echoed by the then Minister of Natural Resources, himself a woodlot owner, who adopted the term "special gift" in an address to assembled forest stakeholders. He suggested that the gift may take time to unwrap but expressed his unwavering solidarity in seeing that appropriate action is taken (MacLellan and Duinker 2012b).

To build something new and vibrant in Nova Scotia will not be straightforward. An ongoing commitment to community-based forest management will challenge many preconceptions about values, ideology, and practice. The province, so tightly hemmed in by the sea and its own history, must act consciously to see the world as it is, and perhaps not as we all might like it to be. Brazil's disastrous flirtation with a local-only approach to forest management has left a legacy of fishbone deforestation and worldwide condemnation. Nova

Scotia's size may avert an ecological disaster on a Brazilian scale and may even contribute to a subtle inclination toward local schemes. But to suggest that the local is *always* better, or at the very least always meaningful, underpins the very nexus of the local trap itself (Cummins 2007). It is here at the highest levels of policy development that understanding of the local trap can best contribute to the avoidance of it.

Finding common ground among stakeholders and achieving compromise will require a healthy dose of humility. The province is beginning from a position of ecological weakness, but Bullock and Hanna (2012) illustrate that community forestry always starts with a degraded land base, socially inoperable setting, and/or decimated regional/local economies. Real progress will take decades, and a community forest network will be a development program not unlike those that sustain forest economies in developing countries worldwide. Nova Scotia's forest sector needs community forests in order to make positive change and progress. As the forest sector was described by Minister Charlie Parker: "This is not a sunset industry. Indeed, the sun is just rising for community forests here in Nova Scotia." (MacLellan and Duinker 2012b: 5). It may be that only future generations will reap the benefits that a reinvigorated forest sector will provide. This should not be interpreted as a disincentive: it is the most altruistic of human motivations.

References

[BCCFA] British Columbia Community Forests Association. 2015. "Status of Community Forestry in BC." Accessed 16 June 2015. http://www.bccfa.ca/index.php/about-community-forestry/status.

Born, B., and M. Purcell. 2006. "Avoiding the Local Trap Scale and Food Systems in Planning Research." *Journal of Planning Education and Research* 26 (2): 195–207.

Brown, C.J., and M. Purcell. 2005. "There's Nothing Inherent about Scale: Political Ecology, the Local Trap, and the Politics of Development in the Brazilian Amazon." *Geoforum* 36 (5): 607–24.

Bullock, R., and K. Hanna. 2007. "Community Forestry: Mitigating or Creating Conflict in British Columbia?" *Society and Natural Resources* 21 (1): 77–85.

———. 2012. *Community Forestry: Local Values, Conflict and Forest Governance.* Cambridge: Cambridge University Press.

[CBC] Domet, S. (Performer). 2012. Interview Charlie Parker. *Mainstreet.* Canadian Broadcasting Corporation, Halifax. Accessed 6 December 2012. http://www.cbc.ca/player/Radio/LocalShows/Maritimes/Mainstreet.

Chappell, C. and J. Simpson. 2010. "A Silvicultural and Economic Comparison of Clearcutting and Partial Cutting Studies in Northeastern North

America." http://www.ecologyaction.ca/content/clearcutting-and-government%E2%80%99snatural-resource-strategy-where-we%E2%80%99re.

Clancy, P. 2001. "Atlantic Canada: The Politics of Private and Public Forestry." In *Canadian Forest Policy: Adapting to Change*, edited by M. Howlett, 205–36. Toronto: University of Toronto Press.

Conrad, V. 2012. "Natural Resources – Bowater Land Sales: Provincial Asset – Ensure." Accessed 8 December 2012. http://buybackmersey.ca/.

Cummins, S. 2007. "Commentary: Investigating Neighbourhood Effects on Health—Avoiding the 'Local Trap.'" *International Journal of Epidemiology* 36 (2): 355–57.

Duinker, P.N., and L.K. MacLellan. 2012. *In Support of Community Forests: Recommendations to the Government of Nova Scotia.* Stewiacke: Nova Forest Alliance and Halifax: School for Resource and Environmental Studies, Dalhousie University.

Duinker, P.N., P. Matakala, F. Chege, W. Patrick, and L. Bouthillier. 1994. "Community Forests in Canada: An Overview." *Forestry Chronicle* 70 (6): 711–20.

England's Community Forests. 2005. "About England's Community Forests." Accessed 6 December 2012. http://www.communityforest.org.uk/aboutenglandsforests.htm.

Fearnside, P.M. 1989. *A Ocupação Humana de Rondônia: Impactos, Limites e Planejamento.* Brasilia: Programa Polonoroeste.

Goldsmith, F.B. 1980. "An Evaluation of a Forest Resource – A Case Study from Nova Scotia." *Journal of Environmental Management* 10 (1): 83–100.

Government of Nova Scotia. 2011. *The Path We Share: A Natural Resources Strategy for Nova Scotia.* Halifax: Department of Natural Resources.

Leger, D. 2011. "The Crisis in our Forests: What's Really Going On." *Chronicle Herald,* 19 September. Accessed 30 July 2014. http://thechronicleherald.ca/opinion/22089-crisis-our-forests-what's-really-going.

Lutzenberger, J. 1985. "The World Bank's POLONOROESTE Project – A Social and Environmental Catastrophe." *Ecologist* 15 (1/2): 69–72.

MacLellan, L.K., and P.N. Duinker. 2012a. *Community Forests: A Discussion Paper for Nova Scotians.* Stewiacke: Nova Forest Alliance and Halifax: School for Resource and Environmental Studies, Dalhousie University.

———. 2012b. *Advancing the Conversation on Community Forests in Nova Scotia: Proceedings from the June 2012 Forum on Community Forests.* Stewiacke: Nova Forest Alliance and Halifax: School for Resource and Environmental Studies, Dalhousie University.

Malhi, Y., J.T. Roberts, R.A. Betts, T.J. Killeen, W. Li, and C.A. Nobre. 2008. "Climate Change, Deforestation, and the Fate of the Amazon." *Science* 319 (5860): 169–72.

Nepstad, D., B.S. Soares-Filho, F. Merry, A. Lima, P. Moutinho, J. Carter, and O. Stella. 2009. "The End of Deforestation in the Brazilian Amazon." *Science* 326 (5958): 1350–51.

[NSDNR] Nova Scotia Department of Natural Resources. 2012. "Community Forests." Accessed 8 March 2015. http://novascotia.ca/natr/forestry/community-forest/.

[NS] Nova Scotia Hansard. 2012. "Debates and Proceedings." Accessed 8 March 2015. http://nslegislature.ca/index.php/proceedings/hansard/C89/house_12nov13/.

Oh, Ho-Sung. 1986. "Economic Development and Changing Forest Problems and Policies: The Case of Korea." In *Community Forestry: Lessons from Case Studies in Asia and the Pacific Region*, edited by Y.S. Rao, M.W. Hoskins, N.T. Vergara and C.P. Castro, 135–47. Bangkok: FAO and East-West Center.

Pagdee, A., Y.-su Kim, and P.J. Daugherty. 2006. "What Makes Community Forest Management Successful: A Meta-study from Community Forests Throughout the World." *Society & Natural Resources* 19 (1): 33–52.

Salonius, P. 2007. "Silvicultural Discipline to Maintain Acadian Forest Resilience." *Northern Journal of Applied Forestry* 24 (2): 91–97.

Sandberg, L.A., and P. Clancy. 1996. "Property Rights, Small Woodlot Owners and Forest Management in Nova Scotia." *Journal of Canadian Studies* 31 (1): 25–47.

Teitelbaum, S. 2014. "Criteria and Indicators for the Assessment of Community Forestry Outcomes: A Comparative Analysis from Canada." *Journal of Environmental Management* 132: 257–67.

Teitelbaum, S., T. Beckley, and S. Nadeau. 2006. "A National Portrait of Community Forestry on Public Land in Canada." *Forestry Chronicle* 82 (3): 416–28.

[UNFAO] United Nations Food and Agriculture Organization. 2012. "Community Forestry." Accessed 8 March 2015. http://www.fao.org/docrep/u5610e/u5610e04.htm.

United Nations Department of Economic and Social Affairs. 1992. *Rio Declaration on Environment and Development*. Rio de Janeiro. Report of the United Nations Conference on Environment and Development, 3–14. Accessed 16 June 2015. http://www.un.org/documents/ga/conf151/aconf15126-4.htm.

Community Forestry on the Cusp of Reality in New Brunswick

Tracy Glynn

Deep in the heart of the New Brunswick forest in a place where a mighty river flows is a dream of a community forest. Upper Miramichi, better known as Boiestown and by the names of the smaller communities around it, like many New Brunswick communities, has historically been and continues to be dependent on the forest for employment and its community well-being. Following a wave of mill closures in the area that had employed many, a forward-thinking mayor, already known for his willingness to abandon the status quo and try something different, flirted with the idea of starting a community forest within his community's municipal boundaries of Upper Miramichi.

Scott Clowater became the first mayor of Upper Miramichi in 2008. Clowater, a retired federal government employee and firefighter, was a driving force behind incorporating Boiestown and the surrounding communities into a rural municipality that then gave the area some rights to manage the Crown land within their municipality's borders. Upper Miramichi transitioned from a local service district managed by the province to an incorporated municipality providing powers to an elected mayor and council.

Clowater and his team hired Sarah Carson-Pond, who had a background in public relations and human resources, as their administrative clerk. Carson-Pond became immersed in a world of learning about community forests. She contacted the Conservation Council of New Brunswick, a long-time advocate for community forests, to find out more about how community forests operate. Shortly thereafter, on 19 March 2009, the Falls Brook Centre, a sustainable community demonstration and training centre in rural western New Brunswick, with support from the Conservation Council, held a Forest Forum. The Forest Forum gathered community forest practitioners like Chris Caldwell from the Menominee Nation in Wisconsin and Jennifer Gunter from the BC Community Forestry Association and representatives from municipalities and First Nations, as well as woodlot owners, conservationists, students, professors, and others, to discuss issues facing forest-dependent communities. Participants discussed tenure reform as a key ingredient in making community forestry a reality in New Brunswick.

The Conservation Council, Carson-Pond, and municipal representatives saw great potential for a community forest in Upper Miramichi. They envisioned a different model for forestry in New Brunswick than the existing industrial model, one wherein community members would have access to and make management decisions concerning forest resources in their area and make decisions on how forests within or adjacent to community boundaries would be managed. They were inspired by the growing number of examples of working community forests in British Columbia that demonstrated that community forests have the potential to diversify local economies, restore and protect forest ecosystems, and provide a more equitable distribution of wealth from forest resources (see Gunter and Mulkey, Chapter 8; also Egunyu and Reed, Chapter 9; and Leslie, Chapter 10).

The first thing the partnership of municipal and non-government organizations did was to form the New Brunswick Community Forest Alliance. Inspired by the Northern Ontario Sustainable Communities Partnership (see Palmer and Smith, Chapter 2), the alliance drafted a community forest charter and gathered traditional and non-traditional allies together to discuss community forestry in New Brunswick. Alliance meetings were attended by woodlot owners, First Nations foresters, mayors, and long-time conservationists. A constituency of support was built for a community forest pilot project, and community groups were called upon to share experiences and support the growing network.

For example, through ongoing alliance meetings, Steven Ginnish from Eel Ground First Nation shared experiences of the history of community forestry in his community with those who were newly interested in building such initiatives in their own communities. The Eel Ground Community Forest had generated employment and operated a sawmill in the 1990s. Ginnish noted that the economic potential for a community forest was constrained by the limited land base and the degraded state of the forest (see also Betts [1997] for further details). The community forest no longer operates in Eel Ground, but Ginnish hopes that it will be revived someday.

In addition to local organizing, the alliance took on a policy advocacy role. For example, alliance members met with Department of Natural Resources staff to ask for support for a community forest pilot project. The department staff expressed their willingness to learn more about community forests. The NB Community Forest Alliance, wanting to learn more about community forestry across the border, organized a panel discussion on 29 November 2010, in Fredericton. Approximately eighty people attended the event, including community forest practitioners, conservation group representatives, woodlot owners, municipal employees, Department of Natural Resources staff, and a local Member of the Legislative Assembly. Speakers included Marcy Lyman with the New Hampshire–based Community Forestry Collaborative and Kris Hoffman with the Amherst Mountains Community Forest in Maine. The event provided an essential space for different groups to come together and discuss local involvement in forest management and development.

In the coming years, the alliance continued to solicit input from community forestry experts in order to develop local capacity for involvement. For example, in the summer of 2012, Jennifer Gunter of the BC Community Forestry Association, and a native of Fredericton, was asked to give a presentation to Upper Miramichi residents about the growing community forest movement in British Columbia and different models for successful community forests in BC. Dr. Tom Beckley, a sociologist in the Faculty of Forestry & Environmental Management at the nearby University of New Brunswick in Fredericton, who has studied community forests in Canada, and Ron Smith, a local expert on non-timber forest products, have also acted as resource persons informing the Upper Miramichi community forest effort.

Maps were generated with the support of the Conservation Council and Fundy Model Forest to illustrate the features of Upper Miramichi's forest. The maps were important visual tools that enabled residents to discuss economic opportunities found in their forest, such as fiddlehead, mushroom, and maple syrup production and ecotourism.[1] A colourful pamphlet and survey with beautiful pictures from Upper Miramichi were sent to all 980 households in Upper Miramichi in June 2012. The pamphlet and survey served to create awareness for community forestry and to engage the community on the topic of a community forest in their backyard in the context of locally relevant issues and priorities. Respondents shared their desire to protect the Miramichi River and explore the possibility of promoting ecotourism and the economic development of their region's diverse forest products.

Efforts were made to promote the region's existing forest enterprises and expertise. For example, Graham Lyons's fiddlehead production in Upper Miramichi was featured in a YouTube promotional video.[2] Lyons calls the forest his office, and he thinks that a community forest is a fine idea. Forest festivals were organized annually by the Central Woodsmen's Museum and the Rural Community of Upper Miramichi, with support from the Conservation Council, for three years from 2011 to 2013. The last two festivals were held in conjunction with lumberjack competitions and were attended by hundreds of people. The festivals showcased the variety of non-timber forest products in the area and provided information about community forestry. The community forestry initiative was greatly supported by these and other local examples of leadership, organizing, and capacity development.

A steering committee for a community forest in Upper Miramichi was formed in 2011. Many members on the steering committee were preoccupied that summer with the presence of a company interested in shale gas exploration in their community. They opposed the exploration and saw community forestry as a wiser alternative for the future of their community. They held town hall meetings to convey their concerns with shale gas throughout Upper Miramichi during the fall of 2011. The Conservation Council was invited to present information about community forestry at these town hall meetings. Community forestry had become part of local debate concerning economic development and environmental protection.

The future of Upper Miramichi and many other small communities in New Brunswick will continue to be shaped by the evolving political economy of provincial forest policy. Public debate is ongoing regarding complex issues such as the provincial tenure system and the relative roles of industrial licence holders and communities, as well as public preferences for forest conservation. On one hand, a provincial move toward industrial licence consolidation (from 84 before 1982 to 6 by 2014), coupled with a controversial additional annual allowable cut rate of 660,000 cubic metres in 2014, appears to be putting more control in the hands of existing large firms, namely J.D. Irving, Limited. On the other hand, there is evidence of a strong public desire to conserve New Brunswick's remaining native mixed-wood Acadian forest (Nadeau et al. 2008). There are also communities such as Upper Miramichi that want more control over local forests and First Nations and traditional councils that are saying that the government has failed to consult them about how the forests are managed on Crown land, land that was never ceded by the Indigenous people in New Brunswick. For example, responses from a survey by Wyatt et al. (2015)

done in thirteen of the province's fifteen First Nations communities found that governance arrangements do not deliver the priorities of environmental protection that are important to First Nations and that the power in these arrangements rests solidly with government and industry. The authors conclude that Indigenous rights do not necessarily enable First Nations to access benefits and that the governance arrangements do not provide a guarantee of sustainable forest management.

Forest governance in New Brunswick has long been characterized as a public-private arrangement between government and industry with no defined policy-making process (Ashton and Anderson 2005). New Brunswick's Department of Natural Resources, recently reconfigured and renamed the Department of Energy and Resource Development, is criticized for its delegation of various management responsibilities to the private forestry sector. Some suggest that conservation groups and First Nations occupy the back seat in this arrangement, which has been increasingly challenged for denying a suite of public priorities (Ashton and Anderson 2005; Nadeau et al. 2008).

The legitimacy of forestry governance and practice in New Brunswick has been perhaps most notably challenged by a number of court rulings that affirm treaty rights. A provincial court ruled in 1997 that Reginald Paul, a Wolastoq (Maliseet) man, and the Wolastoq and Mi'kmaq peoples of the province of New Brunswick have a treaty right to harvest timber from Crown land. The case was later overturned in an appeal court, but the court case did push the provincial government to provide limited harvesting rights to First Nations through First Nation Harvesting Agreements (FNHAs). However, decisions over the volume and location of timber to be harvested as well as the price of timber and revenue to be generated from a timber harvest in such arrangements are not made by the First Nations but by government and industry (Wyatt et al. 2015).

Seven Mi'kmaq First Nation Chiefs announced on 12 May 2016 that they are suing the New Brunswick government over the 2014 forestry strategy, claiming that the strategy and agreements signed between government and the forestry companies infringe on Aboriginal and treaty rights of the Mi'kmaq. The statement of claim argues that the forestry strategy will have a permanent and negative impact on wildlife and the overall health of the forests of New Brunswick, and will adversely affect Mi'kmaq rights to hunt deer and moose, fish salmon and trout, and gather in the forest. The chiefs argue that the government did not meaningfully consult with them before signing the agreements. A temporary court injunction was unsuccessfully sought by some First Nations in 2014 to stop the implementation of the strategy set for February 2015. The Government of New Brunswick has stated that it will honour the contracts with the forestry companies and that they will review the 2014 forestry strategy. However, the government has yet to announce any changes to the previous government's forestry plan.

Critics of the forestry strategy, including conservation groups, First Nations, woodlot owners and smaller mill owners, are particularly concerned that the increased softwood fibre supply granted to the forestry industry, mostly to J.D. Irving, will come from lands set aside for conservation, will further harm the province's already struggling woodlot owners, and will shut out other economic opportunities such as community forestry. Another group of professionals quickly mobilized to call for a halt to the new forestry strategy when it was made known to the public: 184 professors at the province's four public universities and the Maritime College of Forest Technology sent an open letter to then Minister of Natural Resources Paul Robichaud on 16 May 2014. The professors pointed out the lack of citizen input in the plan and noted the negative implications that the plan would have for wildlife conservation.

When finalized, the 2014 forestry strategy shocked those engaged in forest policy debates. A decision that had been delayed for over decade finally granted industry's desire for an increased and guaranteed softwood fibre supply from Crown lands. A number of reports commissioned by the government and industry were completed to add to the forest policy-making discussion. However, there were few public consultations on the reports over the previous decade. The bilateral negotiation approach used by industry and government for Crown land and forest management—the norm over the past 100 years (Ashton and Anderson 2005)—persists in New Brunswick today as exemplified by the way the 2014 forestry strategy was handled.

Industry has long fought for a timber objective in the overarching vision statement for Crown lands management. They commissioned, with the government, the 2002 report *New Brunswick Crown Forests: Assessment of Stewardship and Management*, by consulting firm Jaakko Pöyry. The report contained the controversial recommendation of doubling the annual allowable cut on Crown land over fifty years. The report was part of industry's response to the establishment of ten new protected natural areas on Crown lands. The forestry companies argued that the withdrawal of forested lands from their forestry licences in the amount of 150,000 hectares would have adverse economic impacts and lead to job losses. An all-party committee of Members of the New Brunswick Legislative Assembly (MLAs), the Select Committee on Wood Supply, was formed in light of the public outcry to the Jaakko Pöyry report. The select committee held thirteen public hearings across the province (Ashton and Anderson 2005). The message heard at well-attended public hearings in 2003 and 2004 was the need to assign wood allocations to communities when mills close, reduce clearcutting, and protect and restore the diversity of the forest. The select committee rejected the proposal to double the annual allowable cut on Crown land. Over a decade later, the province's chief financial watchdog reminded MLAs of the numerous studies and recommendations that have called for a reduction of clearcutting in Crown forests. New Brunswick's Auditor General Kim MacPherson reported in 2015 that 80 percent of all the wood cut from Crown forests in the past two decades had been harvested by clearcutting.

The public hearings of the Select Committee on Wood Supply connected and mobilized people across the province concerned with the trajectory of Crown forest management. The Crown Lands Network, through the facilitation of the New Brunswick Environmental Network, formed and included the participation of the Conservation Council, the New Brunswick chapter of the Canadian Parks and Wilderness Society, and local forest watch groups across the province. The groups worked for over a decade on a variety of shared concerns, including clearcutting and herbicide spraying and what they saw as alternatives, namely, community forestry. The groups believed that the Crown Lands and Forest Act enacted in 1980 was outdated and in need of modernizing so that it could reflect the public priorities for forest management, including optimizing local benefits from the use of local resources. A newly formed coalition of organizations calling for modernized Crown forest legislation went public in late 2015 with a press conference in downtown Fredericton. The Conservation Council, the Canadian Parks and Wilderness Society, and the New Brunswick Federation of Woodlot Owners, long-time advocates of modernized forest legislation, joined together with those new to the call, the New Brunswick Wildlife Federation, the New Brunswick Salmon Council, and Nature NB. Community forest proponents, including those engaged in the Upper Miramichi community forest effort, agree with the growing coalition calling for modernized forest legislation. They reference the experiences in British Columbia, where modernized forest legislation, part of which allowed communities access to a sufficient amount of the resource, was key to enabling community forests to get established and multiply in that province.

With such contentious issues commanding public and government attention, community forestry remains what optimists could call an emerging opportunity in need of constant advocacy in order to take hold. In this context, the Upper Miramichi experience demonstrates the importance of local leadership and collective action for building capacities that communities need to become more involved in forest management and economies. By remaining engaged with communities, advocates and knowledge holders, a growing network, informational tools, and shared values were developed to provide essential infrastructure needed to pursue a community forestry agenda in New Brunswick.

Notes

1 These maps are available for public viewing at the Upper Miramichi municipal office or online at https://uppermiramichicommunityforest.wordpress.com/.

2 See https://www.youtube.com/watch?v=k6EhnhoKoTQ.

References

Ashton, B., and B. Anderson. 2005. "New Brunswick's 'Jaakko Pöyry' Report: Perceptions of Senior Forestry Officials about Its Influence on Forest Policy." *Forestry Chronicle* 81 (1): 81–87.

Betts, M. 1997. "Community Forestry in New Brunswick." *International Journal of Ecoforestry* 12 (3): 247–54.

Nadeau, S., T.M. Beckley, E. Huddart Kennedy, B.L. McFarlane, and S. Wyatt. 2008. *Public Views on Forest Management in New Brunswick: Report from a Provincial Survey.* Fredericton: Natural Resources Canada, Canadian Forest Service, Atlantic Forestry Centre.

Wyatt, S., M. Kessels, and F. van Laerhoven. 2015. "Indigenous Peoples' Expectations for Forestry in New Brunswick: Are Rights Enough?" *Society & Natural Resources* 28 (6): 625–40.

Fostering Community Capacity, Enterprise, and Diversification

The British Columbia Community Forest Association: Realizing Strength in Regional Networking

Jennifer Gunter and Susan Mulkey

The British Columbia Community Forest Association (BCCFA) is a network of rural community-based organizations engaged in community forest management, and includes communities seeking to establish new community forests. Over nearly two decades, community forestry has taken root in British Columbia (BC), and the BCCFA has worked to help create a healthy and vibrant community forest sector.

Forests cover about 60 percent of BC's total land mass. These forests are richly diverse, and have been classified into sixteen distinct biogeoclimatic zones. Ninety-five percent of the land in the province is publicly owned (BCMFML 2010). This public ownership has made space for the creation of the community forest agreement tenure on Crown land.

The following is a brief overview of the development of community forestry in BC and the role that the BC Community Forest Association has played in the success of this innovative form of forest tenure.

History and Formation of Community Forestry in British Columbia

In 1998, the Province of BC introduced the Community Forest Pilot Program. The program was the culmination of decades of public action to create opportunities for greater community and First Nation participation in the management of local forests. Gunter (2000) earlier described this evolution. The concept of community forestry first developed in British Columbia in the 1940s when Gordon Sloan in the Royal Commission on the Forest Resources of BC of 1945 (Sloan 1945) recommended that municipalities manage local forests. This recommendation led to the establishment of the Mission Municipal Forest. The second Sloan Commission recommended expanding this concept to involve other municipalities, but nothing came of this proposal (Burda 1999).

In the 1970s and 1980s, as public awareness of the need to protect forest ecosystems from the negative impacts of industrial logging practices began to grow, the idea of community forestry grew in favour. Water and soil quality, fish and wildlife habitat, and wilderness preservation all became issues of public concern. While environmentalists were focusing their attention on preservation, a growing number of people who worked with communities and First Nations, and in the labour movement, were becoming equally concerned with responsible management of the "working forest" (Pinkerton 1993). Yet another royal commission led in 1976 by Peter Pearse supported the expansion of community forests. He said, "Local governments that are prepared to integrate their lands with surrounding Crown forest land is one attractive possibility. The sensitive balance between timber production, recreation, and other non-commercial forest and uses that are particularly valuable close to centres of population can in these cases be struck locally, making resource management highly responsive to local demands" (Pearse 1976: 118).

Subsequently, several concepts were discussed to promote the devolution of decision-making authority to communities (see Pinkerton 1993; Tester 1992; Maki et al. 1993; Burda et al. 1997); however, it was not until the creation of the community forest pilot agreement that community forestry took hold.

In conjunction with the 1997 Jobs and Timber Accord, a major initiative by the BC Ministry of Forests to create new employment and economic opportunities in the forest sector, volume was made available to communities (BCMF 1997). In 1998 the Forest Act was amended to include a new form of tenure, the community forest agreement. Communities were invited to submit applications in a competitive process for a new experimental community forest pilot program. Nearly ninety communities expressed interest in the pilot program and twenty-seven submitted full applications. By 2001, ten community forest pilot sites had been identified. The new pilots, with their five-year probationary licences, joined the ranks of a handful of BC communities that were already managing traditional forest tenures with a community forestry mandate.

In 2003, a second major government initiative was introduced, the Forestry Revitalization Plan, which brought with it changes to the Forest Act and new volume and area available for the community forest program. Thirty-three additional communities were invited to participate in the program. Then in

March 2009, the BCCFA's advocacy efforts influenced changes to legislation that removed the probationary aspect of the tenure, making community forest agreement holders eligible to secure a long-term renewable licence, thereby improving the incentive to invest in planning and business relationships.

British Columbia's Community Forest Program

The provincial government's (BCMFLNRO 2014) goals for the Community Forest Program, as they directly appear in stated provincial policy, are to:

1. provide long-term opportunities for achieving a range of community objectives, values, and priorities;
2. diversify the use of and benefits derived from the community forest agreement area;
3. provide social and economic benefits to BC;
4. undertake community forestry consistent with sound principles of environmental stewardship that reflect a broad spectrum of values;
5. promote community involvement and participation;
6. promote communication and strengthen relationships between Aboriginal and non-Aboriginal communities and persons;
7. foster innovation; and
8. advocate forest worker safety.

Community Forest Agreements (CFAs) are primarily timber tenures granted to legal entities that represent a local community's interests. CFAs can be held by a municipality, community corporation, co-operative, society, First Nation band council, or partnership. The agreements are area-based and grant the holders exclusive rights to harvest timber. The tenure also grants non-exclusive rights to harvest, manage, and charge fees for botanical forest products and other products, as well as the ability to manage for water, recreation, wildlife, and viewscapes (see Leslie, Chapter 10 and Egunyu and Reed, Chapter 9, for practical examples). CFA holders have stewardship responsibilities that include strategic and operational planning, inventory maintenance, and reforestation, along with obligations to engage with and report to the community. Community forests are subject to the provisions of the Forest and Range Practices Act, and agreement holders pay stumpage fees to the Crown based on a tabular rate structure. CFAs are issued for a 25-year term, and are replaceable every 10 years. The existing tenures range in size from 1,081 hectares to 160,212 hectares, with an average size of 27,000 hectares.

To understand community forests in BC, it is important to recognize the key elements of the CFA tenure. As described by Leslie (2016), BC's community forests have forest tenure rights that, while limited in scope, are relatively strong and secure. The tenure is area-based, long-term, and not transferable. These characteristics create incentives for long-term planning and investments in the health and productivity of the forest. These incentives can be understood in the context of property rights, as described by Bullock, Teitelbaum, and Lawler in Chapter 2 of this volume.

Research in many natural resource sectors has indicated that the more complete the set of rights held by an individual or group, the more likely they are to develop rules that define how they exercise their rights of withdrawal (i.e., timber harvesting) (Schlager and Ostrom 1992). The incentives to develop management regimes that are sustainable and avoid overexploitation are stronger when users are faced with the long-term consequences of their decisions. In contrast to other forms of forest tenure in BC, CFA holders must truly face the long-term consequences of their decisions—and conversely are motivated to invest for future benefit. This is because the CFA, unlike other industrial forest tenures, does not confer the right to alienate. In this context, this right refers to a licensee's ability to sell or transfer their tenure.

The CFA and the new First Nations Woodlands Licence are the only tenures in BC that are, in effect, not transferable. In the case of the CFA, the licensee does not have the right to "alienate" or sell its tenure. The only transfer that can occur is to another legal entity representing the community in question (for example, from a community-based Society to a Co-op), and only with the approval of the Minister of Forests, Lands and Natural Resource Operations. This is a fundamentally important distinction to be made, especially when looking at what incentives exist to invest in enhanced forest stewardship and the future economic value of the forest.

The Strength of a Network

The initiation of the new community forest pilots quickly revealed numerous challenges. Provincial forest policy that was designed for large industrial companies, not for the new small-scale community tenures, was identified early on as an impediment to success. In addition, there was a strong desire on the part of those involved in the pilot program, along with other community leaders with an interest in community forestry, to share their experiences and to learn from one another.

By 2002 there was broad recognition that it was time to form a province-wide organization and create a collective voice for community forests. With just ten inaugural member communities, the BC Community Forest Association (BCCFA) was formed. It was conceived as a grassroots, inclusive, and member-driven organization. In 2014, representing over fifty communities in BC, the BCCFA has preserved its original intent, and has become the voice of BC communities engaged in community forest management as well as those seeking to establish community forests.

The BCCFA is a legally incorporated not-for-profit society open to all organizations and individuals who support the vision, mission, purposes, and guiding principles of the association. Membership is voluntary, and is available to all existing community forest organizations in BC, including those that have forest tenure, as well as community forest organizations that are seeking to obtain local forest management rights. The BCCFA's membership comprises a wide array of communities, but the population of each member community is under 20,000, with the majority under 10,000. About one third of the membership is First Nation–held community forests or a partnership involving First Nations. Full members of the association have the right to elect directors and to vote on key decisions. In addition, non-voting associate and supplier memberships are available to individuals and organizations that support the vision, mission, purposes, and guiding principles of the association.

A nine-member volunteer board of directors is elected by the full members and operates as a policy governance board whose role is to set policy and strategic priorities. BCCFA contract staff implements the strategic plan and manages the day-to-day operations of the association. The core operations of the BCCFA are funded through membership dues. Special projects are funded through outside sources and partnerships that include the federal and provincial governments, regional trusts, development organizations, and private foundations.

The Role of the BCCFA

The mission of the BCCFA is to promote and support the practice and expansion of sustainable community forest management in BC. The mandate of the BCCFA is focused in three key areas: advocacy, networking, and education.

The BCCFA serves as the conduit between the expressed interests and needs of the membership and government decision makers, actively working to secure improvements to the legislation, regulation, and policy that affect community forests. To this end, the BCCFA has established a mutually beneficial relationship with the provincial government with a focus on improving the program and facilitating the success of community forest initiatives. The association's work is rooted in the core belief that community forests are the best tenure option for public land around rural communities.

The organization has grown in step with the expansion of the community forest program in the province, and significant strides have been made to enable the success of the tenure. For example, through the efforts of the BCCFA, in 2006 CFAs were recognized by government as a unique form of tenure and a new pricing arrangement was put in place.

Another primary activity of the association is the provision of support to communities in BC who are striving to obtain and successfully implement community forest initiatives. Through the network of practitioners, in conjunction with partners in academia, industry, government, and the non-profit sector, the BCCFA works to increase awareness of the unique approaches required in community-based forest management and to provide practical information on best practices drawn from experience. The strong collaborative network supports communities with resources to increase their organizational capacity, community engagement, and transparency, and to share information on forest management and marketing of forest products.

Educational activities focus on member priorities and include an extension program in partnership with the University of British Columbia (UBC) Faculty of Forestry. The program offers in-community meetings where community forest practitioners can air their specific questions and challenges concerning community forest governance and forest management. An annual conference, monthly newsletters, and educational publications further serve to deepen the understanding of community forest management, and support networking and organizational development. Since the introduction of the community forest tenure in BC, many valuable lessons have been learned.

In an effort to document these lessons and to make tools available to practitioners in BC, the BCCFA has published two Community Forestry Guidebooks in partnership with FORREX (Gunter 2004; Mulkey and Day 2012).

Challenges and Lessons

Many of the challenges that community forestry in BC faces are internal—rooted in governance at the community level—while others are external in nature and are the result of the public policy context within which the community forest tenure resides, along with general market trends as well as forest health. The BCCFA works with communities to bring these challenges to light and to facilitate the generation of options that will best suit each unique community.

Since the community forests are governed by volunteer boards, local capacity to establish effective organizational governance structures is vital. There is an ongoing need for education and training to understand and navigate the complexities of community processes and organizational governance issues related to the management of a community resource. For example, a common stumbling block is how to distribute the benefits derived from the community forest. A key principle of community forestry is that the people in the community share the benefits obtained from the forest within which they live. The profits (including non-monetary benefits) stay in the community and are widely distributed. Experience shows that policies for fair distribution of benefits must be established and openly communicated, preferably before financial benefits have accrued.

The goals of community forests include generating local jobs and economic, social, and ecological resilience in rural areas. To meet these aspirations, solid business planning must be the foundation of community forest efforts. Starting with small successes, building capacity over time and ensuring a secure financial bottom line is essential for a community forest organization. However, the majority of community forests in BC are small operations and are constrained by economies of scale. This is especially challenging at a time when industry corporate consolidation is growing. The BCCFA is actively advocating for an increase in the size of existing community forests to improve their long-term economic viability.

Community forests can play a role in maximizing the value of the forest resources under their control through innovation and by seeking synergies in new partnerships. Catalyzing regional economic linkages can be a core management objective for community forest organizations, as elaborated by Bullock, Teitelbaum, and Lawler, Chapter 1, and Lachance, Chapter 5. CFA holders are in a position to support local entrepreneurship through access to long-term timber supply and local agreements. Alliances with strategic partners in the public and private sector can further expand opportunities to add value. The BCCFA is actively promoting regional dialogue and development of regional economic clusters with an eye to the emergence of new economic opportunities.

Political Support: An Essential Requirement

Since the introduction of the CFA, interest and support has grown continuously in BC and goes beyond a single political party. The original pilots were established by an NDP government; the BC Liberals expanded the program and have worked to streamline the administrative requirements. Ongoing support for the tenure is clearly stated in the province's forest sector reports and strategies.

The Working Roundtable on Forestry (2009: 7) articulated a vision for a "vibrant, sustainable, globally competitive forest industry that provides enormous benefits for current and future generations and for strong communities." This included the key priority of supporting prosperous rural forest economies, with the recommendation that "we should expand the Community Forest Tenure Program."

Building on the Working Roundtable on Forestry's recommendations, the potential of community forests is further acknowledged in the province's Forest Sector Strategy (BCMFLNRO 2012: 21) with its plan to "improve access to forest tenure for a range of users," with the commitment: "New community forest agreement opportunities will be created where suitable areas and fibre supplies exist or where partnerships can be created."

The BC experience clearly shows that political support is essential to increase community participation in sustainable forest management. Secure land tenure and effective policy frameworks are central to the continued growth of community forestry as a viable feature on the BC forest management landscape. The major milestones regarding the shaping and improvement of the community forest program have come about through the co-operative relationship between the BCCFA and provincial and local government representatives. This relationship continues to grow as mutual successes are achieved.

While there is widespread support for community forestry, the further growth of the program is uncertain. The establishment of new CFAs and the expansion of existing ones are challenged by the fact that there are many competing interests on the land base. Most important are unsettled Aboriginal land claims that the BC government is working to resolve. Some First Nations have acquired CFAs as part of their negotiation process, yet not all First Nations have access to this option. The majority of the province's annual timber harvest is already allocated to other forest tenures. The lack of unallocated harvest volume is compounded by the reality of a decreasing wood supply resulting from the mountain pine beetle epidemic. Further gains for community forests will require collaboration on the part of the BCCFA, individual communities, First Nations, and local, regional, and provincial governments, along with the support of the broader forest industry.

Conclusion

The decade-and-a-half of the community forest tenure in BC has clarified a number of factors which enable their establishment and sustainability. While the story behind each community forest is unique, the successes to date in BC can be attributed to some common themes identified by Gunter (2000):

- It is essential that there be secure tenure and clear management rights. Forest management policy needs to accommodate the unique characteristics of community-based management. This requires political support at all jurisdictional levels.
- Local capacity and the development of skills in organizational governance, particularly volunteer board development and transparent community engagement, are vital. Communities must access financial resources for startup, be able to operate the community forest as an economically sound business, and determine early on how they will fairly distribute the benefits back to the community.
- The establishment and tending of a network of community forest practitioners that openly share information and experiences and collaborate with other organizations will advance the community forest movement. Development of a strong collective voice with a mandate to speak for member communities will support effective advocacy.

Within Canada, BC is a leader in community forestry. The recent introduction of the First Nations Woodlands Licence, a forest tenure for First Nations based on the CFA framework, will create even more opportunities for long-term community-based management. That being said, the scope of the management rights conferred by the CFA, with its focus on timber, is limited. A number of communities want to manage the lands that surround them in a more integrated and holistic manner, beyond what can be achieved within the legislative and regulatory framework of the CFA tenure. As community forestry takes root and grows in different forms in other jurisdictions in Canada, there are likely to be new ideas and insights that can inform its future development in BC.

References

[BCMF] British Columbia Ministry of Forests. 1997. "Communities to Play Greater Role in Forest Management—Zirnhelt." Press Release on 3 December. Accessed 18 June 2015. http://www2.news.gov.bc.ca/archive/pre2001/1997/1997nr/1997102.asp.

[BCMFLNRO] British Columbia Ministry of Forests, Lands and Natural Resource Operations. 2014. "Government Objectives for Community Forest Agreements." Accessed 18 June 2015. https://www.for.gov.bc.ca/hth/timber-tenures/community/objectives.htm.

———. 2012. *Our Natural Advantage: Forest Sector Strategy for British Columbia*. Victoria: Queen's Printer. https://www.for.gov.bc.ca/mof/forestsectorstrategy/Forest_Strategy_WEB.PDF.

[BCMFML] British Columbia Ministry of Forests, Mines and Lands. 2010. *The State of British Columbia's Forests*, 3rd ed. Victoria: Forest Practices and Investment Branch. www.for.gov.bc.ca/hfp/sof/index.htm#2010_report.

Burda, C. 1999. "Ecosystem-based Community Forestry in British Columbia: An Examination of the Need and Opportunity for Policy Reform, Integrating Lessons from around the World." Master's thesis, University of Victoria.

Burda, C., D. Curran, F. Gale, and M. M'Gonigle. 1997. *Forests in Trust: Reforming British Columbia's Forest Tenure System for Ecosystem and Community Health*. Victoria: Eco-Research Chair of Environmental Law and Policy, Faculty of Law and Environmental Studies Program, University of Victoria.

Gunter, J. 2000. "Creating the Conditions for Sustainable Community Forestry in BC: A Case Study of the Kaslo and District Community Forest." Unpublished Master's Research Project, School of Resource and Environmental Management, Simon Fraser University.

Gunter, J., ed. 2004. *The Community Forestry Guidebook: Tools and Techniques for Communities in British Columbia*. FORREX Series Report No. 15. Kamloops: FORREX – Forest Research Extension Partnership, and Kaslo: British Columbia Community Forest Association. http://www.bccfa.ca/index.php/what-we-do/publications/item/89-guidebook.

Leslie, E. 2016. "Stronger Rights, Novel Outcomes: Why Community Forests Need More Control over Forest Management." In *Community Forestry in Canada: Lessons from Policy and Practice*, edited by S. Teitelbaum, 311–28. Vancouver: University of British Columbia Press.

Maki, T., G. Walter, and S. Hutcheson. 1993. *Community Sustainability and Forest Resource Use: Discussions with Community Leaders in the Alberni-Clayoquot and the Cowichan Valley Regional Districts*. Victoria: Sustainable Communities Initiative Component Three Working Group, Centre for Sustainable Regional Development, University of Victoria.

Mulkey, S., and J.K. Day, eds. 2012. *The Community Forestry Guidebook II: Effective Governance and Forest Management*. FORREX Series Report No. 30. Kamloops: FORREX – Forum for Research and Extension in Natural Resources, and Kaslo: British Columbia Community Forest Association. http://bccfa.ca/wp-content/uploads/2013/03/FS30_web-proof.pdf.

Pearse, P. 1976. *Timber Rights and Forest Policy in British Columbia.* Report of the Royal Commission on Forest Resources. Victoria: Queen's Printer.

Pinkerton, E. 1993. "Co-Management Efforts as Social Movements: The Tin Wis Coalition and the Drive for Forest Practices Legislation in B.C." *Alternatives* 19 (3): 34–38.

Schlager, E., and E. Ostrom. 1992. "Property-Rights Regimes and Natural Resources." *Land Economics* 68 (3): 249–62.

Sloan, G. 1945. *Report of the Commissioner Relating to the Forest Resources of British Columbia.* Victoria: C.F. Banfield, King's Printer. http://www.llbc.leg.bc.ca/public/pubdocs/bcdocs2011/274554/report%20of%20the%20commissioner,%20forest%20resources%20of%20bc,%201945.pdf.

Tester, F. 1992. "Reflections on Tin Wis: Environmentalism and the Evolution of Citizen Participation in Canada." *Alternatives* 19 (1): 34–41.

Working Roundtable on Forestry. 2009. "Moving Toward a High Value, Globally Competitive, Sustainable Forest Industry." British Columbia. https://www.for.gov.bc.ca/hfd/library/documents/bib108842.pdf.

Harrop-Procter Community Forest: Learning How to Manage Forest Resources at the Community Level

Felicitas Egunyu and Maureen G. Reed

This chapter reports on a study that investigated the contributions of social learning to forest governance in Harrop-Procter Community Forest in British Columbia (BC), Canada. It provides an example of research about social learning in an operating community forest in BC and in so doing contributes to a better understanding of community forestry practice and innovations. It begins with the history of Harrop-Procter Community Forest (HPCF). It then continues with a description of community forestry in Canada and how social learning may occur within community forest settings. There is a brief overview of data collection methods after which research findings are organized around three key activities where learning occurred. The chapter then assesses the effects of learning on HPCF's practices and ends by considering the effects of learning for resource management practice and governance arrangements.

History of Harrop-Procter Community Forest

Harrop-Procter Community Forest is located outside the communities of Harrop and Procter about thirty kilometres northeast of Nelson in the West Kootenay region of southeastern BC (see Figure 9.1). The community forest land base covers 11,300 hectares of provincial forest Crown land on the south shore of the west arm of Kootenay Lake. The communities of Harrop and Procter have a population of 650 people (Statistics Canada 2012). In addition to being rather small, they are also isolated in that the only way to reach the

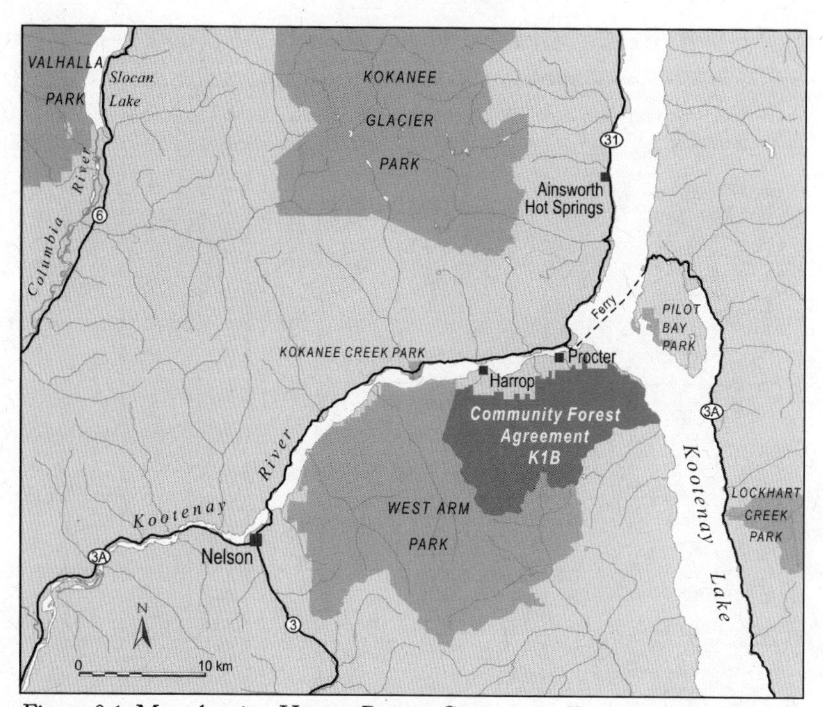

Figure 9.1. Map showing Harrop-Procter Community Forest and nearby communities.

communities is by cable ferry. Although there is no First Nation presence in the communities, HPCF falls within the hunting grounds of the Sinixt Nation.

HPCF lies within the Lake Kootenay Timber Supply Area. However, prior to the establishment of the community forest, the forests surrounding the two communities had not been logged for almost two centuries. When the BC Ministry of Forests Lands and Natural Resource Operations (MFLNRO, referred to here as Ministry of Forests) announced plans to log the area in 1985, some people from the communities of Harrop and Procter pressed for the protection of the forests through the formation of a provincial park. The West Arm Provincial Park was formed in 1995, but it excluded Harrop-Procter's forests. Beginning in 1996, the people then lobbied for the establishment of a community forest.

The signing of a Community Forest Agreement (CFA) was the culmination of a decades-long process in which most of the local people fought to have direct control over what happened in their forests. The government awarded a five-year licence with an allowable annual cut (AAC) of 2,600 cubic metres in July 1999. In 2013, the AAC was increased to 10,000 cubic metres at the request of HPCF. The Ministry of Forests abolished five-year CFAs in 2008 and HPCF received a twenty-five-year licence which is renewable every ten years.

HPCF is overseen by the Harrop-Procter Watershed Protection Society (the Society) and the Harrop-Procter Community Co-operative (the Co-op). The Society's role is stewardship, and the Society and Co-op's role is management. See Table 9.1 for additional information about each organization. Some board members sit on both boards to provide linkage between the two organizations.

Table 9.1. Description of Harrop-Procter Watershed Protection Society and Harrop-Procter Community Co-operative.

Name	**Harrop-Procter Watershed Protection Society**	**Harrop-Procter Community Co-operative**
Year formed	1996	1999
No. of board members	9	11
No. of shared board members	5	5
No. of members (March 2014)	37	137
Mandate	Forest & watershed protection; ecosystem-based forest development; employment; research & public education	Holds CFA; runs forest management business; employment
Membership	Open to all people	Residents of Harrop & Procter
Board meetings	Quarterly	Monthly
Member meetings	Annual, non-members welcome	Annual, non-members welcome
Membership fees/ categories	Annual/ residents: $10 Individual, $25 family Annual/non-resident $10 Life/resident: $100	Life/residents/individual: $25
Revenue source	Membership fees, bequests, grants, fundraising	Membership fees, bequests, business income

(Source: HPCF)

The Society started as a committee in 1985 and registered in 1996. It submitted the proposal for the community forest in 1998. At the time of the submission of the HPCF proposal, the Society had 276 members—half of the total population of both villages. The Society is overseen by a board of directors that is elected from the membership. The Co-op was formed in 1999 to hold the CFA and run forest operations as well as pursue economic development. Harrop-Procter chose the co-op model because it allowed public participation and maintained accountability to the community (HPCF 2014). Unlike the Society, membership in the Co-op is for life and is restricted to residents of Harrop and Procter. As of March 2014, the Co-op had 137 members, most of whom are also members of the Society. The Co-op is also overseen by a board of directors that is elected from the membership. Nine of the eleven board members were interviewed for the study.

HPCF's forest management philosophy is influenced by Silva Forest Foundation's[1] ecosystem-based conservation planning model. All the wood harvested in the community forest is certified by the Forest Stewardship Council (FSC).[2] The Silva Foundation and the FSC promote management approaches that require learning by practitioners.

Community Forestry in British Columbia

In Canada, community forestry is implemented under provincial, territorial, or municipal forestry legislation (Teitelbaum et al. 2006; Bullock et al. Chapter 1). The diversity of settings gives rise to different types of community forests, mostly on public provincial land, but some are also located on land owned by counties and municipalities (Bullock et al., Chapter 1). As outlined by Gunter and Mulkey, Chapter 8, BC established a community-based forestry tenure to enable community forestry on Crown land. Community forestry is described by the Province as "any forestry operation managed by a local government, community group, or First Nation for the benefit of the entire community" (MFLNRO 2014).

The Ministry of Forests, through the Forest Act, provides guidance on forest management and operations as well as the level and timing of community involvement. Importantly, community involvement starts before the signing of the CFA with the local community forest organization. The Ministry of Forests requires documented evidence of community involvement in the community forest proposal, and community involvement is expected to continue throughout the tenure of the community forest. Community consultations are required for forest operations management planning, including timber harvest volume revisions.

Local communities managing a community forest must adhere to government regulations irrespective of their experience with forest management or available resources, or their CFA can be cancelled. Local communities that hold CFAs usually face challenges that include limited human and financial resources, external threats to forest ecosystems, and working with limited knowledge (Reed and McIlveen 2006; Bullock et al. 2009). Few community forest volunteers begin with a professional background in forestry. The challenges that community forest organizations face require learning about forest management, appropriate governance, day-to-day operations, and strategic decision making, among other things. As economic, ecological, and social conditions are also subject to change, communities engaged in forestry also have to learn to adapt during management if the community forest is to continue to function.

Social Learning in Community Forest Settings

Community forestry is a form of collaborative governance that has become popular in developed countries since the 1990s (McCarthy 2006; Charnley and Poe 2007). It has been variously defined, but it generally involves managing forests for social, environmental, and economic values within a government policy framework (Teitelbaum et al. 2006; Bullock and Hanna 2012). Community forests are increasingly used as collaborative decision-making models in various parts of the world, including Canadian forestry-based communities. Community forestry is participatory in nature, attracting people with different interests, needs, values, resources, and knowledge levels. People usually work together to manage forestry resources for values that have been agreed upon by the local community. However, the values for which a community forest will be managed are also either set or strongly influenced by government (McCarthy 2006; Coulibaly-Lingani et al. 2011). Hence community forests in Canada, for example, are usually managed for timber harvesting because the government policy framework sets them up that way, whereas in Africa and parts of Asia they are usually managed for firewood and non-timber forest products (NTFPs).

Community forest practice, like other environmental governance arrangements, can be enabled by social learning (e.g., Cheng et al. 2011; Fernandez-Gimenez et al. 2008; Bullock et al. 2012). For this study, social learning is defined as "a change in understanding that goes beyond the individual to become situated within wider social units or communities of practice through social interactions between actors within social networks" (Reed et al. 2010: n.pag.). In addition to learning that occurs through transformative community organizing (see Palmer and Smith, Chapter 2), social learning occurs as people participate in resource management activities such as meetings, data collection, advocacy, awareness raising, field tours, fundraising, resource harvesting, and monitoring. As people learn to work together, they can alter governance processes, for example, by generating opportunities for greater collaboration among stakeholders with different values, needs, and skills (Biedenweg and Monroe 2013; Fernandez-Gimenez et al. 2008; Bull et al. 2008; Reed et al. 2010). Learning sometimes results in community-initiated collective action for sustainable forest resource management (Rist et al. 2007; Biedenweg and Monroe 2013). Other times learning results in support for conservation initiatives (Sinclair et al. 2011) or collective understanding about other values and needs (Cheng and Mattor 2010; Leys and Vanclay 2011). Researchers generally agree that learning leads to more adaptive and enduring socio-ecological systems (Cheng et al. 2011; Berkes 2009). Learning can also take place through critical reflection on activities (Marschke and Sinclair 2009; Cundill 2010).

Community forestry researchers have examined specific characteristics of community forests such as the level of power devolution (Ambus and Hoberg 2011); how power relations affect management (Reed and McIlveen 2006); conflict management (Bullock and Hanna 2007; Bullock et al. 2009); community forest models (Bixler 2014); ingredients for success (Pagdee et al. 2006); and criteria for assessment (Teitelbaum 2014). But there is still a need to understand how social learning, community forest management activities, and governance processes are interconnected. This chapter contributes to a better understanding of how social learning influences management activities within a community forest, particularly within a developed country setting.

Research Approach

The study was qualitative, using a case-study approach. Data were collected from June to July 2013 using personal interviews, focus group meetings, and participant observation. Twenty-eight people were interviewed from Harrop-Procter villages: fourteen were male, and fourteen were female. Three of the twenty-eight interviewees were not members of either the Society or the Co-op. Of the remaining twenty-five interviewees, eleven were current board directors. Two focus group meetings were held with three and eight interview participants respectively. Three forest walks were also conducted to get a better understand of ecological and social values associated with the community forest. Additionally, six interviews were conducted with employees of HPCF, the British Columbia Community Forest Association, and current and former employees of the Ministry of Forests. Data were analyzed following a mix of inductive and deductive approaches and the qualitative analysis program NVivo was used to assist in coding and drawing out themes.

Research Findings

Community Forest Activities as Sites for Social Learning

What people learn when directly involved in community forestry depends on the activities they engage in and the values, knowledge, needs, and expectations they bring to the community forest. HPCF was formed out of a desire for local control over logging within forests in the community. As a participant explained, "Whether we thought it was a good idea or not there was going to be logging occurring here and that was the impetus for us to say, 'Okay we want a community forest,' and then log on our specifications rather than the standard forestry [male participant]." Although the founding members of HPCF were interested in watershed protection and community forest management, they did not have backgrounds in community forest management or co-operative business management. They learned as they interacted with each other and external partners in managing their forest.

Interviewees described learning while carrying out various projects that HPCF implemented. For example, when the Co-op started a sawmill, people described learning about mill set-up, operations, and product sales. When timber harvest volumes were revised, people reported learning about sustainable forestry. Interviewees admitted to starting with a limited knowledge and to learning as they got involved in HPCF: "When I started I didn't understand very much at all what was going on but by participating you learn [male participant]."

Without participation there is almost no learning. People described different activities that they participated in and how they learned: "I learned it through the allowable annual cut process [female participant]"; and "I have learned just from going to meetings and public meetings . . . I learned from [forest manager's] presentations recently when we did a new management plan for the forest [female participant]." The set-up of HPCF provided ample opportunities for members and interested non-members to learn: "The nature of a co-op is that it is a co-operative venture, all the members theoretically participating in it. . . . So it is a mutual education experience and by and large that's the theme that runs through the management and the relationship between our board of directors and the members [male participant]." So by mutually engaging in various community forest activities, members learned. Table 9.2 provides additional examples of what interviewees reported learning at Harrop-Procter. The examples are arranged according to variables identified from published literature on natural resource management and social learning.

Table 9.2. Social learning variables from published literature and outcomes identified from interviews with residents of Harrop-Procter villages.

Variables from published literature	Social learning outcome examples from personal interviews at HPCF
Knowledge	Forest ecology, community forest management, watershed function and protection, co-operative/business management, board governance, mill operations, fundraising
Technical skills	Forest monitoring, water quality monitoring, wildlife surveys and counts, timber cruising, business management
Social skills	Leadership skills, communication skills
Action	Closure of a non-profitable project (Sunshine Bay Botanicals)
Relationships	Starting to develop good working relationship with Ministry of Forests
Shared understanding	Shared understanding of: sustainable community forest business, forest ecology, forest management, need for increased timber harvest volumes (AAC)
Values & attitudes	Changed attitudes about loggers, now that they too are logging, changed assumptions about "high-volume" logging
Organization structure	Changed board structure to increase efficiency Developed new organization (Harrop-Procter Business Products) to process and sell forest products.

(Sources of variables: Schusler et al. 2003; Rist et al. 2007; Fernandez-Gimenez et al. 2008; Brummel et al. 2010; Biedenweg and Monroe 2013)

From the above table we see a variety of learning outcomes at individual (knowledge and skill acquisition) and group levels (development of shared purpose or change in organization structure). These learning outcomes all contributed to the success of HPCF as a well-managed community forest.

Learning to "Do" Community Forestry

To illustrate how HPCF evolved as members learned and implemented community forestry, three examples of major changes at the community forest are examined. These changes highlight what was needed, learned, and implemented as well as outcomes of social learning processes. The changes are:

- the evolution of Harrop-Procter Community Co-operative board of directors
- the story of Sunshine Bay Botanicals
- increase in timber harvest volumes.

Evolution of Harrop-Procter Community Co-operative Board of Directors

The first example of social learning for effective governance is the evolution of the Co-op board. The Co-op holds the CFA and manages the forestry business. The Co-op is overseen by a board of directors; each board member serves a two-year renewable term. As of 2013, the board had five female and six male directors.

Interviewees reported that the Co-op board had changed in terms of processes, capacity, and composition. With regard to processes, by 2011 the board became more efficient than it was at the outset of the community forest in 1999. For example, board meetings had become shorter, lasting on average two instead of four hours. This occurred since the board changed its procedures; it now circulates reports prior to board meetings, has a meeting chairman to move discussions based on the agenda, and has subcommittees to handle issues and report to the board. These strategies reduced the length of board meetings and also increased the amount of work done by board members between meetings. As one board member described, "That was part of the restructuring, trying to figure out, okay, we don't want these meetings to go on forever. So how can we cram as much productivity in two hours of a board meeting? And how do we have to communicate and handle information? So now we send everything out that can be sent prior, we have a chair that actually runs the meeting with an agenda that is circulated prior [male participant]." Another board member described other strategies: "We've only just decided the last couple of years to have committees, which is a few board members who are joined together to do a particular thing. And they change quite a bit. We have a finance committee, we have an environment committee, we have a membership committee, and it's usually a couple of board members who will do it twice a year, something like that. They are responsible for something that needs doing which we bring up in the meeting, they will take it on [female participant]."

While changes to the board's activities may seem routine and ordinary, they represent a learning-based approach (variables: knowledge, organization structure in Table 9.2). The board's capacity also changed; there are now people with more technical background and experience in forestry and forestry-related business on the board than was the case when HPCF was established. The new procedures and capacity have enabled the board to be more efficient and effective.

Board composition also changed so that board members are now more "homogenous," to borrow a participant's word. The participant noted, "the reason we have been successful in the last few years is because it is a very, very homogenous group . . . we have the same or similar values and when we tackle an issue, we usually have good discussions [male participant]."

From this interviewee's description, it appears the board has developed a shared purpose and understanding with regard to HPCF management (variable: shared purpose/understanding in Table 9.2). But it also appeared that people with different opinions/values about board governance or timber harvest volumes, for example, have stopped volunteering for the board. As one interviewee explained, "I quit from the Co-op because my input was not effective . . . when I was on the Co-op committee I felt completely ineffectual. Because my voice was such a minority that I was achieving nothing: I just couldn't jump on the band wagon, the agenda that was presented and everybody else did basically [male participant]." This interviewee later described the Co-op's objectives as more developmental than the HPCF's original objectives.

The board composition has also changed from "homogenous" people who wanted to devote "energy to the cause" to "homogenous" people who offered specific technical expertise. Some former board members reported stepping down because they felt they did not have the technical expertise. For example, one member shared that "after I had been involved for about three or four years, what I could see, they needed a specific skill set, you know. I'm not really good at computers and stuff like that and fundraising, I don't have those skills. And so, I thought well, I just felt like I wasn't useful and so I resigned from the committee. You know, I felt that was the right thing to do [female participant]."

It should be noted that board positions are open to every interested HPCF member but certain types of people (e.g., people who support low timber harvesting levels, people who do not have specific technical skills, people with young families, etc.,) are now opting not to volunteer.

Learning to Let Go: The Story of Sunshine Bay Botanicals

In addition to harvesting timber, one of HPCF's objectives was to manage non-timber forest products (NTFPs). Sunshine Bay Botanicals was established in 2001 as a division of Harrop-Procter Community Co-operative to grow, harvest, make, and sell NTFPs. Members of the Co-operative grew and harvested herbs; they also harvested wild herbs and medicinal plants and made crafts (e.g., potato boxes), tinctures, and herbal teas. The project was staffed by local volunteers. In addition to volunteers, the company used students who were paid by funds earmarked for such activities from external sources such as the Columbia Basin Trust. Sunshine Bay Botanicals was an innovative business in that it did what other BC community forest businesses did not do—that is, harvest, process, and sell NTFPs.

For some years, Sunshine Bay Botanicals enjoyed some success. It was recognized by community forest and community co-operative proponents as a truly innovative community business (Krause and Faust 2001; BCICS 2001). However, Sunshine Botanicals was closed in 2007 once logging got underway and the mill was established.

The response to the project was mixed. Some members viewed it as innovative and successful. But other interviewees described the project as one that used up a lot of volunteer time and organizational resources and yet was not profitable. They thought it was too much work and they were satisfied that it was closed. Other participants saw the project as a "filler project." For example, an interviewee described it as "something kind of to bridge the gap until we could start logging and we could start road building. So it was like the beginning part [female participant]."

It served a purpose, giving the community forest organization something to do and earn money while other activities were set up. It also provided volunteer opportunities for people who were not interested in timber products. But other interviewees wished it could be taken up again, either because they missed the products or because it was a project they felt they could volunteer with. As an interviewee explained, "I think it is hard when you are learning how to operate a community forest and you are learning how to run a mill and you are learning how to log and . . . how to sell and how to market, and you are going round teaching courses or doing talking events . . . maybe it was a little bit much. So maybe now that they have, they kind of have the logging down and the community has settled down a little bit, maybe they could try the value added stuff [e.g., teas, tinctures, etc.] again [female participant]."

One main lesson from the Sunshine Bay Botanicals was learning how to take on projects that could be handled within the available resources (variables: technical skills and action from Table 9.2). As one participant summed it up, "So there's this kind of limit to growth . . . it was a lesson in thinking we could do more than was possible not because the ideas were bad but because we didn't have the money, the people, the time or expertise to pull it off all at the same time [male participant]."

When the Society first applied for a community forest licence, the HPCF's proposal was seen as being competitive and different from other community forest proposals and successful because, apart from logging, it also proposed the harvest and processing of NTFPs and tourism. HPCF had to let go of one of its main businesses (Sunshine Bay Botanicals) because it used up a lot of resources in an already resource-strained organization that needed to concentrate on Harrop-Procter Forest Products to make it viable. Letting go of a project is an important part of learning, as it indicates the ability to recognize what is working and needed at a particular time within the life of the organization. But the loss of Sunshine Botanicals effectively narrowed the scope of activities and learning opportunities under the community forest licence.

Cutting More Trees to Protect Community Watersheds

The final major change associated with learning is the increase in the allowable annual cut (AAC). When submitting a community forest proposal, the AAC is negotiated with the Ministry of Forests based on the available AAC in the region. The ministry has no recommended AAC for community forests, but it is generally agreed that the AAC should be large enough to enable the community forest to be financially viable as well as provide benefits to the local community.

When HPCF was awarded the licence in 1999, their AAC was 2600 cubic metres. This harvest volume was one of the lowest allowed by the province; HPCF was more interested in protecting their watersheds and preserving their forests than in logging. About eight years later, there was a temporary increase in the AAC to get rid of beetle-infested lodgepole pine (*Pinus contorta*). From 2012 to 2013, HPCF reviewed their AAC and consulted with members as well as the general community about a potential AAC increase. In 2013, with support from members and the local community, the new AAC was set at 10,000 cubic metres. Even at 10,000 cubic metres, this remains one of the lowest AACs in British Columbia. According to HPCF's own analysis, they can set their AAC at 10,000 cubic metres and harvest sustainably for another forty years (HPCF 2014).

There was support for the increased AAC within the HPCF membership and the villages of Harrop and Procter as shown by the results of a community survey as well as from this study's interview respondents. However, of the twenty-eight study participants who were interviewed, five felt the new AAC was too high given that HPCF started with 2,600 cubic metres. Other participants felt that raising the cut was okay but only up to a level beyond which a value line is crossed; for some people that line had already been crossed with the new higher AAC.

Some community forest members had reservations about a higher cut; they reasoned that if operational costs were kept low, if fewer people were employed, and if capital developments were reduced, then a higher AAC would not be justified. Others said they needed the high AAC in order to run a sustainable community forest business and retain direct management of their forests. As one person summed it up, "we need to cut more trees in order to protect our water [female participant]."

The female participant reasoned that it was possible to harvest more trees, be profitable, and still do it in a way that protected the environment and ensured that the area would still be forested for future generations. This view was also reiterated by a number of study participants, some of whom reported learning during the AAC revision process (variable: knowledge in Table 9.2). Watershed protection is a concept that is very dear to Harrop-Procter inhabitants because they obtain their drinking water from the creeks in their forested watersheds. Careless logging—which they had experienced before on private land—could affect drinking water quality and quantity. But in order to be able to retain management of logging operations, they needed funding, and since timber harvesting is their major source of revenue, they needed to cut more trees to protect their water.

Interviewees stressed that successful watershed protection was linked to a financially sustainable community forest logging business. As one participant explained:

> We are in the business of running a business and I have said to others what we are doing is this. Number one, we are looking after our water supply because everybody, including those who work in the community forest, need[s] water, okay. Once we settle on how we are going to protect our supply, which is what we have done, we then said, alright let's go logging. That is a bit different than the way things work in traditional forestry in British Columbia. First we look after our water, then we log. And then all the logging we do, of course, is done in keeping with the preservation of our water systems. That said, once we have embarked on this community forest idea, we are a business. We have debt to pay, we have overhead to pay, we have people to employ. We have an operation that has to be accounted for to the community. We have to do it in as businesslike a way as we amateurs can possibly do. [Male participant]

Figure 9.2. HPCF-donated wood used for the benches in their local Sunshine Bay Park and for the wharf in the background.

Proponents of the new higher cut argued that they were still managing the community forest based on the founding mandates and values. They also maintained that they hold Forest Stewardship Council certification to prove their dedication to sustainable forest management. As an interviewee explained, "we have maintained eco-certification, so that's a huge thing to say that we are doing a good job with managing the forest for certain values [female participant]."

For others, the certification was to assure members that the community forest had not left its founding values and mandate: "Certification is an assurance to the community that we are doing it the best that we can do it. I think that gets way more money value than how much extra [dollars] we can tuck onto [the price of] a stick of wood [female participant]."

Over the years, as HPCF learned the community forest business, members learned that they needed to harvest more timber in order to be sustainable as a business. According to the majority of interviewees, cutting more trees provided much-needed funds for community forest activities, including logging operations. It also enabled HPCF to pay stumpage fees, carry out capital investments, provide local employment, run a sustainable business, donate lumber to local community initiatives (e.g., donated wood to build a children's play park, a community wharf, and park benches—see Figure 9.2 below), and ultimately protect its water sources. But harvesting at four times the original cut may indicate a shift in values that is not supported by some HPCF members, especially some founding members who were more forest preservation–minded. These values may be closer to local conventional forestry than are some of the broader ideals of community forestry. But HPCF is not the only community forest in BC to increase timber harvest volumes; Logan Lake Community Forest also increased its AAC from 20,000 cubic metres to 100,000 cubic metres.

Effects of Learning on Community Forestry Practice

We see HPCF evolving as members learned and practised community forestry. This evolution was enabled by the social learning that occurred during the various activities of the community forest. For example, the people who participated in HPCF's activities reported that they acquired the necessary knowledge, social, and technical skills to perform activities that enabled them to meet the government's community forest requirements. Learning became situated within a broader setting; for instance, we see the Co-op board of directors learning and improving their output as they worked toward reducing time wastage.

HPCF, like other community forests in BC, faced challenges such as limited financial and human resources. Members learned to work creatively with the resources they had and reorganized the board so that they could get more done. Additionally, members had to learn to prioritize businesses and let go of other initiatives like Sunshine Bay Botanicals that were not performing as expected. On the other hand, learning outcomes resulted in a narrowing of focus, reducing opportunities for ongoing learning in a different area, such as the procurement of NTFPs. Indeed, there were interviewees who had previously volunteered with the NTFPs project and when it ceased operation admitted to not having much else to do beside board directorship or attending meetings. When Sunshine Bay Botanicals was shut down, the community forest lost a number of volunteers and ongoing learning opportunities.

Other researchers have found that social learning enables community forest organizations not only to acquire and build up the knowledge and skills needed for forestry practice but also to build skills for participants to actively engage in community forest-based or forest-initiated activities (e.g., Fernandez-Gimenez et al. 2008; Biedenweg and Monroe 2013). We found this to be true for HPCF; the interviewees described acquiring knowledge and skills related to community forest management that they previously did not have. In addition, community forest organizations can also experiment with different management approaches. Hence, community forest organizations can opt for different governance configurations; for example, they can use a corporate structure like Creston Community Forest in BC (Bullock and Hanna 2012), or a co-operative structure like HPCF.

One of the Ministry of Forests community forest objectives is to "foster innovation" (MFLNRO 2014). HPCF was also able to try out innovative practices with regard to timber and NTFP harvesting, so that instead of harvesting only timber using conventional commercial clearcut harvesting methods, they tried other harvesting methods, such as selective cutting. However, this research also found that innovative practices can be costly to a community forest in terms of both financial and human resources. In one case, HPCF tried cultivating, harvesting, and selling NTFPs through Sunshine Bay Botanicals. They were the first community forest in British Columbia to harvest and sell NTFPs. The business ran for six years through heavy volunteer and external funding support. As discussed by Murphy, Chretien, and Morin (Chapter 11), labour and supply costs associated with NTFP production and processing can be quite high. Consequently, HPCF closed the business because it used up more human and financial resources than they could afford, and this allowed the community forest to concentrate on developing the timber value-added business, which brought in much-needed funds for forestry operations.

Conclusions

Community forestry has become popular as a form of forest governance in Canada only in the last decade or so. Given that community forest implementation is guided by existing legislation and moderated by local context, there is no one structure that fits everywhere. Rather, governance is collaborative, with specific community forest mandates and practices guided by local context. And because of this, researchers and practitioners are still trying to understand community forestry in practice. This chapter joins other publications in contributing to that end.

HPCF, like most community forests in BC, is largely run by volunteers. Volunteers are drawn from the local community and come with different skill levels as well as different values. Most HPCF volunteers did not begin with community forest management skills but have learned as they participated in various activities and implemented their CFA. Our research at HPCF confirmed that community forest organizations do provide a place for volunteers to acquire and practise knowledge, technical, and social skills.

Social learning was found to contribute to improved governance at HPCF; learning enabled participants to streamline the board and increase timber output. HPCF learned how to implement community forestry while using minimal resources; they also learned to stop activities that were draining their limited financial and human resources and enlarged their values to obtain more money from their community forest harvests.

This study of HFCF also highlights the tension between innovation and organizational survival. HPCF tried to be innovative in some of its activities; for example, in the types of businesses it operates by selling NTFPs. However, innovation in providing NTFPs in HPCF's case, even with external financial support, had a high financial and human cost and could not be sustained. BC's Ministry of Forests may desire to foster innovation among community forest organizations, but the cost of innovation may be prohibitive to community forest organizations. Community forest organizations may have to select between innovation and profitability.

A surprising finding of the study is that success at community forest business could lead to a reduction of volunteer avenues. It would be interesting to examine other community forest organizations to investigate whether community volunteer opportunities become reduced as a community forest becomes successful financially. Most BC community forest organizations started out as volunteer organizations. Along the way, some have developed a more businesslike approach and have replaced volunteers with hired workers. Pressure to run community forests like a business have also been felt at Wetzin'kwa Community Forest in Smithers, BC. Even though this community forest has generated local profits for distribution to the community, it has struggled to engage general community members in its activities and decision-making processes (Assuah et al. 2016). Do community forest organizations that have demonstrated financial success provide fewer avenues for general community participation or volunteering? Addressing this question in future research will shed light on how to achieve balance between two competing demands—to use community forests as vehicles for direct public engagement in management of public resources and to use community forests as vehicles for enhancing local community profitability.

Whether or not a community forest organization sets out deliberately to learn, its members will learn. The tension, long evident in community-based resource management, between learning to become a viable business and learning how to maintain local involvement remains in relation to community forestry (Bradshaw 2003; Reed and McIlveen 2006; McIlveen and Bradshaw

2009). Anticipating social learning and deliberately providing room for self-examination and reflection after the implementation of activities will enable social learning outcomes to inform subsequent community forest activities. By highlighting social learning processes and outcomes within a community forest context, this chapter helps uncover both the strengths and limitations of community forestry in practice, considers the potential and realization of the capacity of community forests to innovate, and introduces new questions for ongoing research.

Notes

1 Silva Forest Foundation is a non-profit BC organization that champions the use of ecosystem-based conservation planning.

2 The Forest Stewardship Council is an international not-for-profit organization that promotes the sustainable management of forests around the world.

References

Ambus, L., and G. Hoberg. 2011. "The Evolution of Devolution: A Critical Analysis of the Community Forestry Agreement in British Columbia." *Society and Natural Resources* 24 (9): 933–50.

Assuah, A., J. Sinclair, and M.G. Reed. 2016. "Action on Sustainable Forest Management through Community Forestry: The Case of the Wetzin'kwa Community Forest Corporation." *Forestry Chronicle* 92 (2): 232–44.

[BCICS] British Columbia Institute for Co-operative Studies. 2001. "Situating Co-operatives in British Columbia." 2000–2001. Victoria.

Berkes, F. 2009. "Evolution of Co-Management: Role of Knowledge Generation, Bridging Organizations and Social Learning." *Journal of Environmental Management* 90: 1692–1702.

Biedenweg, K.A., and M. Monroe. 2013. "Teasing Apart the Details: How Social Learning Can Affect Collective Action in the Bolivian Amazon." *Human Ecology* 41: 239–53.

Bixler, R.P. 2014. "From Community Forest Management to Polycentric Governance: Assessing Evidence from the Bottom Up." *Society and Natural Resources* 27 (2): 155–69.

Bradshaw, B. 2003. "Questioning the Credibility and Capacity of Community-Based Resource Management." *Canadian Geographer* 47 (2): 137–50.

Brummel, R.F., K.C. Nelson, S.G. Souter, P.J. Jakes, and D.R. Williams. 2010. "Social Learning in a Policy-Mandated Collaboration: Community Wildlife Protection Planning in the Eastern United States." *Journal of Environmental Planning and Management* 53 (6): 681–99.

Bull, R., J. Petts, and J. Evans. 2008. "Social Learning from Public Engagement: Dreaming the Impossible?" *Journal of Environmental Planning and Management* 51 (5): 701–16.

Bullock, R., D. Armitage, and B. Mitchell. 2012. "Shadow Networks, Social Learning, and Collaborating through Crisis: Building Resilient Forest-based Communities in Northern Ontario, Canada." In *Collaborative Resilience: Moving through Crisis to Opportunity*, edited by B. Goldstein, 309–37. Cambridge, MA: MIT Press.

Bullock, R., and K. Hanna. 2007. "Community Forestry: Mitigating or Creating Conflict in BC?" *Society and Natural Resources* 22 (1): 77–85.

———. 2012. *Community Forestry*. Cambridge: Cambridge University Press.

Bullock, R., K. Hanna, and S. Slocombe. 2009. "Learning from Community Forest Experience: Challenges and Lessons from British Columbia." *Forestry Chronicle* 85 (2): 293–304.

Charnley, S., and M.R. Poe. 2007. "Community Forestry in Theory and in Practice: Where Are We Now?" *Annual Review of Anthropology* 36: 301–36.

Cheng, A.S., C. Danks, and S.R. Allred. 2011. "The Role of Social and Policy Learning in Changing Forest Governance: An Examination of Community-Based Forestry Initiatives in the U.S." *Forest Policy and Economics* 13: 89–96.

Cheng, A.S., and K.M. Mattor. 2010. "Place-Based Planning as a Platform for Social Learning: Insights from a National Forest Landscape Assessment Process in Western Colorado." *Society and Natural Resources* 23 (5): 385–400.

Coulibaly-Lingani, P., P. Savadogo, M. Tigabu, and P. Oden. 2011. "Factors Influencing People's Participation in the Forest Management Program in Burkina Faso, West Africa." *Forest Policy and Economics* 13: 292–302.

Cundill, G. 2010. "Monitoring Social Learning Processes in Adaptive Co-management: Three Case Studies from South Africa." *Ecology and Society* 15 (3): 28.

Fernandez-Gimenez, M.E., H.L. Ballard, and V.E. Sturtevant. 2008. "Adaptive Management and Social Learning in Collaborative and Community-Based Monitoring: A Study of Five Community-Based Forestry Organizations in the Western USA." *Ecology and Society* 13 (2): 4.

[HPCF] Harrop-Procter Community Forest. 2014. "Harrop-Procter Community Forest." Accessed 7 July 2014. http://www.hpcommunityforest.org/.

Krause, A., and R. Faust. 2001. *North American Case Study: Harrop-Procter Community Forest, British Columbia, Canada: Promoting and Protecting Mountain Products*. Banff: Parks Canada.

Leys, A.J., and J.K. Vanclay. 2011. "Social Learning: A Knowledge and Capacity Building Approach for Adaptive Co-Management of Contested Landscapes." *Land Use Policy* 28 (3): 574–84.

Marschke, M., and A.J. Sinclair. 2009. "Learning for Sustainability: Participatory Management in Cambodian Fishing Villages." *Journal of Environmental Management* 90: 206–16.

McCarthy, J. 2006. "Neoliberalism and Politics of Alternatives: Community Forestry in British Columbia and the United States." *Annals of the Association of American Geographers* 96 (1): 84–104.

McIlveen, K., and B. Bradshaw. 2009. "Community Forestry in British Columbia, Canada: The Role Of Local Community Support And Participation." *Local Environment* 14 (2): 193–205.

[MFLNRO] Ministry of Forests, Land and Natural Resource Operations. 2014. "Ministry of Forests, Lands and Natural Resource Operations." Access date 7 July 2014. http://www.gov.bc.ca/for/.

Pagdee, A., Y. Kim, and P. Daugherty. 2006. "What Makes Community Forest Management Successful? A Meta-Study from Community Forests throughout the World." *Society and Natural Resources* 19: 33–52.

Reed, M., and K. McIlveen. 2006. "Toward a Pluralistic Civic Science?: Assessing Community Forestry." *Society and Natural Resources* 19: 591–607.

Reed, M.S., A.C. Evely, G. Cundill, I. Fazey, J. Glass, A. Laing, J. Newig, B. Parrish, C. Prell, C. Raymond, and L.C. Stringer. 2010. "What Is Social Learning?" *Ecology and Society* 15 (4): n.pag.

Rist, S., M. Chidambaranathan, C. Escobar, U. Wiesmann, and A. Zimmermann. 2007. "Moving from Sustainable Management to Sustainable Governance of Natural Resources: The Role of Social Learning Processes in Rural India, Bolivia, and Mali." *Journal of Rural Studies* 23: 23–37.

Schusler, T.M., D. Decker, and M.J. Pfeffer. 2003. "Social Learning for Collaborative Natural Resource Management." *Society and Natural Resources* 16 (4): 309–26.

Sinclair, A.J., S.A. Collins, and H. Spaling. 2011. "The Role of Participant Learning in Community Conservation in the Arabuko-Sokoke Forest, Kenya." *Conservation and Society* 29 (1): 42–53.

Statistics Canada. 2012. *2011 Census Profile Harrop/Procter, British Columbia (Code 590129) and British Columbia (Code 59) (table).* Catalogue number 98-316-XWE. Ottawa. Released 24 October 2012. Accessed 16 June 2015. http://www12.statcan.gc.ca/census-recensement/2011/dp-pd/prof/index.cfm?Lang=E.

Teitelbaum, S. 2014. "Criteria and Indicators for the Assessment of Community Forestry Outcomes: A Comparative Analysis from Canada." *Journal of Environmental Management* 132: 257–67.

Teitelbaum, S., T. Beckley, and S. Nadeau. 2006. "A National Portrait of Community Forestry on Public Land in Canada." *Forestry Chronicle* 82 (3): 416–28.

Fire and Water: Climate Change Adaptation in the Harrop-Procter Community Forest

Erik Leslie

The Harrop-Procter Community Co-operative (HPCC) manages an 11,300-hectare community forest on Crown land in British Columbia's southern interior. As a local community co-operative, HPCC's mandate comes directly from its membership. Since nearly every household in the Harrop-Procter area obtains its drinking water directly from the streams in the community forest, the primary mandate of the community forest is to protect water. Mature forests, among other things, are very effective in producing a regular supply of clean water. In Harrop-Procter's mountainous terrain, water protection has historically been synonymous with large riparian and headwaters reserves, and a relatively low timber harvest rate.

The vast majority of the forests in Harrop-Procter are approximately 100 years old, having originated from large fires associated with mining and settlement activities in the early twentieth century. In the past fifty years, active forest fire suppression efforts have been quite effective in limiting the impact of wildfire in the area. Until recently, there has also been little logging activity in Harrop-Procter. The Harrop-Procter landscape is thus currently dominated by unbroken tracts of mature coniferous forest. Until recently, this green carpet of mature forest was considered a benevolent asset to be maintained in perpetuity.

When the community forest was created in the late 1990s, climate change was not a major consideration. Natural disturbances such as insect epidemics and fire were briefly discussed in Harrop-Procter's initial management planning but were understood to occur at relatively low rates historically. It was assumed that the "natural" state of Harrop-Procter's forests generally tended toward relatively stable, often "climax" old-growth conditions. Since the objective of forest management was to manage for clean water, and to maintain forest conditions within the historic "range of natural variability," the emphasis was on an extensive network of permanent reserves throughout the community forest. Forest management was designed to safeguard largely undisturbed mature forest conditions and to promote further development of old growth forests.

The Kutetl Wildfire and Emerging Climate Change Challenges

On 8 August 2003, a lightning strike started a wildfire in the headwaters of Kutetl Creek, a remote area just south of the community forest. Several years of drought had led to extreme fire hazard conditions and dozens of fires were already burning across the southern half of the province. The Kutetl area was largely inaccessible, and fire crews were busy fighting higher priority fires in Kelowna and elsewhere, so the Kutetl fire was largely left to burn. After a few weeks the fire had grown to several thousand hectares. Then, contrary to expectations, the fire crossed a mountain ridge and began to burn down into Harrop Creek. By this time, over 100 firefighters had been deployed to fight the fire. However, it was impossible to control the fire without better access, and without rain. Smoke filled the sky and ash was drifting down on the community. Residents of Harrop and Procter were understandably nervous. Evacuation plans were put into place. Since a small cable ferry provides the only vehicle access to Harrop and Procter, emergency egress from the community of 800 residents is highly limited.

By early September the fire had grown to nearly 8,000 hectares, and still there was no rain. Residents could see the orange glow of the fire at night. Large helicopter pads were cut out in Harrop Creek. Heavy equipment was working twenty-four hours a day in an attempt to build a fire guard in the adjacent West Arm Provincial Park. Finally on 8 September, with a community evacuation order imminent, it rained. Eventually the fire intensity abated, and by the end of September the fire was finally extinguished. Luckily, in the end, only about 5 percent of the Harrop Creek watershed was burned.

The Kutetl wildfire was a wake-up call for many residents in the Harrop-Procter area. The high wildfire risk in the area could no longer be ignored; decades of fire suppression had led to a potentially explosive wildfire hazard. Protection from fire became a higher priority forest management objective, and large unbroken tracts of mature coniferous forest were no longer considered benign. There was also a growing recognition in the community of the risk that wildfire presents to water quality in the streams that the community relies on for drinking water. Hydrologists and soil scientists explained that large wildfires can lead to significant soil erosion, stream sedimentation, and disruption of hydrological regimes. A changing understanding of disturbance ecology and wildfire risk presented a challenge to HPCC's initial forest management assumptions and direction.

Climate change projections indicate that summers in the BC southern interior will be considerably hotter and drier. Kootenay region weather station data already shows that over the last fifty years, average temperatures have increased at a rate of over 2°C per century (Pacific Climate Impacts Consortium 2013). It now seems that the 2003 drought and wildfire season was a herald of summers to come. Extreme weather events, once infrequent, are also becoming more common. The rate of forest "disturbances" such as insect and disease epidemics and wildfire is expected to increase significantly (Woods et al. 2010). HPCC now recognizes that many areas of the community forest will not likely develop into historic "climax" conditions. With a changing climate, ecosystem resilience will be severely tested. The community forest will need to adapt to new realities and evolving community priorities.

Practical Steps on the Ground

HPCC recognizes that its forest management regime needs to adapt to the realities of a changing climate, and it has begun to take practical steps to address climate change and wildfire risks. For example, while still largely relying on natural regeneration in many harvest areas, HPCC is also planting seedlings grown from seed collected from slightly warmer and drier areas south of the community forest.[1] HPCC has also begun planting ponderosa pine, a native species well adapted to summer drought and wildfire. On warm, dry, west-facing slopes, HPCC is using regeneration harvest methods to transition existing stands away from cedar and hemlock forests toward primarily Douglas fir and larch forests. While cedar and hemlock are considered climax species at lower elevations in the community forest, they are not very drought tolerant and present a higher fire risk. Douglas fir and larch are better adapted to drier conditions and fire.

In the wildland–urban interface area near private land, HPCC has been working on reducing forest fuels through thinning and the removal of dense cedar-hemlock undergrowth. However, understory fuel treatments are expensive and funding is limited. HPCC is working on using revenue from commercial thinning or small patch cuts to subsidize understory fuel reduction treatments. HPCC is now looking at all potential harvest areas through the lens of reducing fire hazard.

At the landscape level, HPCC has targeted forests at higher risk of drought and wildfire (e.g., forests on dry sites) for commercial logging and has begun breaking up the uniformly dense and mature coniferous forest in these areas. HPCC has developed one modest landscape-level fuel break along a gentle ridge between two drainages. This landscape-level fuel break is a linear 150-metre-wide harvest block that will not be actively regenerated with conifers. Instead, less flammable deciduous shrub growth will be encouraged. More landscape fuel breaks are planned.

HPCC is thus beginning to develop an integrated system of stand- and landscape-level fuel breaks to help protect homes and private property from wildfire. This project is challenging given the community forest's steep terrain, which is often difficult to access. HPCC is not willing to take excessive risks building roads on potentially unstable terrain.[2] Fire management will have to be integrated with HPCC's other forest management objectives. Proactive fire management may also have implications for current and future harvest levels.

When HPCC updated its management plan in 2012, climate change adaptation and wildfire management goals and objectives were added. Significantly, after considerable board and community discussion, HPCC decided to increase its harvest rate, in part to enable more active management of the portion of the land base closest to communities.[3] HPCC and the broader Harrop-Procter community are beginning to understand that, in the dynamic fire-driven ecosystems of the BC southern interior, maintaining large tracts of undisturbed mature forests indefinitely is simply not possible, or even desirable. Uniformly dense, mature forests lack landscape-level diversity (Perry et al. 2011) and are not likely to be resilient to the widespread droughts, wildfires, and insect infestations associated with climate change outbreaks (Woods et al. 2010). At the same time, over half of the community forest—including riparian zones, unstable slopes, caribou habitat, and existing old growth stands—remains in reserves, and moist forests at lower risk of drought are being managed for continuous forest cover. HPCC recognizes that we cannot do the same type of forest management everywhere—we need to diversify the forested landscape and adapt to variable and changing conditions.

Next Steps and Discussion

HPCC's climate change adaptation plans and activities to date represent a modest beginning in a new and uncertain era. HPCC is currently developing a new and comprehensive adaptive management approach and an integrated silvicultural strategy that explicitly address climate change. The adaptation strategy will identify priority areas for conversion to more open and fire-adapted forest types and will provide direction regarding how logging and regeneration activities should be carried out. Cut block size, shape, and orientation will be addressed, and management prescriptions will be developed for each major ecosystem type in the community forest. Multiple harvest rate scenarios will also be explored and publicly discussed.

Since the Kutetl wildfire, and with climate change moving to the forefront of public consciousness, the community discussion in Harrop-Procter regarding environmental change, community priorities, and risk has been evolving. As a grassroots community co-operative, the development of HPCC's climate change adaptation strategy will include deep community engagement over several years. Local residents will be involved directly in discussions of climate change risks and engaged in the forest management decision-making process. Recent public meetings indicate a high level of concern regarding climate change and considerable interest in discussing implications for forest management. Proactive discussions of climate change adaptation thus present an opportunity to renew public engagement in the activities and priorities of the community forest. Climate change adaptation fuels the ongoing social learning process that underpins the evolution of the community forest (see also Egunyu and Reed, Chapter 9).

HPCC's initial work on its climate change adaptation strategy includes outreach to community forests and forestry licensees across southeastern British Columbia. Other Kootenay community forests, including the Slocan Integral Forestry Cooperative, are also actively engaging in climate change adaptation and wildfire management (see www.sifco.ca/integral-forestry/wui). Many community forests across the British Columbia interior are leading efforts to reduce wildland–urban interface fuels and protect communities from wildfire (BCCFA 2016). We are witnessing a shift to a new adaptive forest management paradigm, and community forests, with their broad mandate to respond to new public priorities, are leading the way.

Modern forest management, traditionally focused on a sustained yield of timber, has relied heavily on assumptions of predictable cycles of forest growth and regeneration (Puettmann et al. 2009). When forest ecosystems were assumed to be fundamentally stable over relatively long periods of time, it was possible to imagine that we might have the answers as to how to manage them. Now, the focus is shifting to adaptive management approaches. In an era of great uncertainty and rapid environmental change, it is increasingly important to monitor forest development and be prepared to adapt our management to observed changes. Forest management, now more than ever, requires a wide variety of approaches, an explicit discussion of compromises, and an active balancing of risks. By deeply engaging the public in resource discussions and the decision-making process, community forestry provides unique opportunities for the development of novel, community-based solutions to climate change challenges.

Notes

1 This approach is consistent with new interim provincial seed transfer standards based on "assisted migration." See http://www.for.gov.bc.ca/hfd/pubs/Docs/Tr/Tr048.pdf. Interim BC standards allow seed from lower elevations to be planted further upslope. See also http://www2.gov.bc.ca/gov/topic.page?id=52FDF6763A9D4F278D8EA911FD3EE2AD for a description of the BC Climate-Based Seed Transfer Project.

2 Also, many of the highest priority areas for fuel treatments are on private land, or in the adjacent unroaded West Arm Provincial Park. Effective wildfire protection will require co-operation with private landowners and park managers.

3 HPCC's Management Plan is available at http://www.hpcommunityforest.org/hpcc-management-plan-2012/.

References

[BCCFA] British Columbia Community Forest Association. 2016. *Community Forest Indicators 2015: Measuring the Benefits of Community Forestry*. http://bccfa.ca/wp-content/uploads/2016/02/BCCFA-Report-2015-web2.pdf.

Pacific Climate Impacts Consortium. 2013. *Climate Summary for Kootenay/ Boundary Region*. University of Victoria. http://www.pacificclimate.org/sites/default/files/publications/Climate_Summary-Kootenay-Boundary.pdf.

Perry, D.A., P.F. Hessburg, C.N. Skinner, T.A. Spies, S.L. Stephens, A.H. Taylor, J.F. Franklin, B. McComba, and G. Riegel. 2011. "The Ecology of Mixed Severity Fire Regimes in Washington, Oregon, and Northern California." *Fire Ecology and Management* 262: 703–17.

Puettmann, K.J., K.D. Coates, and C. Messier. 2009. *A Critique of Silviculture: Managing for Complexity*. Washington, DC: Island Press.

Woods, A.J., D. Heppner, H.H. Kope, J. Burleigh, and L. Maclauchlan. 2010. "Forest Health and Climate Change: A British Columbia Perspective." *Forestry Chronicle* 86: 412–22.

Maple Syrup Value Systems and Value Chains: Considering Indigenous and Non-Indigenous Perspectives

Brenda Murphy, Annette Chretien, and Grant Morin

With special thanks to the contributions from Charles Restoule, David Chapeskie, and Melanie Smits

The call to incorporate non-timber forest products (NTFPs) into community forestry is a long-standing one (see Harvey, Chapter 3; also Palmer and Smith, Chapter 2). Community forestry, at least under appropriate conditions, can be a viable alternative to large-scale, industrial, state-run forest management and conventional forms of western forestry. It can better include "multiple knowledge forms and local and non-state actors with different interests and values, as well as consideration of unique local contextual factors" (Bullock and Hanna 2012: 2). In this chapter we argue that community forestry can gain insights about the management and dynamics of NTFPs from a review of maple syrup production. Community forestry projects interested in NTFPs will be faced with many of the same issues as the maple syrup industry as they strive to embrace a range of environmental values, incorporate Indigenous ways of knowing, and focus on long-term sustainability (see Bullock et al. Chapter 1). Through a background review of the syrup industry and the development of two maple value models, in this chapter we provide an overview of the key Ontario geographies, histories, and practices associated with maple syrup production and demonstrate that there are distinctions underpinned by different worldviews and ways of knowing between rural/settler and Indigenous producers. As Casimirri and Kant (Chapter 4) also discuss, understanding these intercultural differences and related power disparities is needed before collaborative undertakings can be effective. Yet, despite these differences, a binary separation between Indigenous and rural/settler producers would underestimate the complex interrelationships that are the hallmark of maple syrup production communities; maple syrup producers are better seen as existing along a continuum. Especially where Indigenous producers blend traditional practices and Indigenous knowledge with recent technological innovations and marketing approaches, producers move toward the centre of this continuum.

Community forestry may not be able to achieve all of its intended goals due to lack of community capacity and resources, entrenched socio-economic relationships and power dynamics, constraining legal and institutional structures, and ecological limitations. And, as will be discussed further below in the context of maple syrup production, additional challenges are presented by the different worldviews among those harvesting the forest resources, particularly Indigenous and rural/settler maple sap and syrup producers. There is also a dearth of information about value-added products, NTFPs, markets, and supply chains (Teitelbaum and Bullock 2013). This chapter, focused on maple syrup production in Ontario, Canada, seeks to provide some insights into these shortfalls. The next two sections provide an overview of the Ontario maple syrup industry. This is followed by the rationale for the development of the two maple models. Subsequent sections outline the research methods, results, and concluding remarks.

Context: Ontario Maple Syrup Production

Commercially, maple syrup is predominantly produced from sugar maples (*acer saccharum*), since the sap from these trees is the sweetest. Sugar maples, with a range that extends across eastern North America (Figure 11.1), are tapped each spring using either a traditional tap and bucket system or the tubing and vacuum pump systems that have been developed since the 1950s. Maple syrup is also used to produce a range of pure maple products such as maple sugar, taffy, and butter, and maple-enhanced products such as sauces, jams, and even spa products. In addition, the sap itself is a natural, nutrient-rich liquid that has always been used as a ceremonial liquid and spring cleanse, especially for Indigenous women (McGregor 2013, Project Advisory Board, personal communication). Sap has recently begun to be marketed commercially as a refreshing beverage, similar to the recent emergence of coconut water.

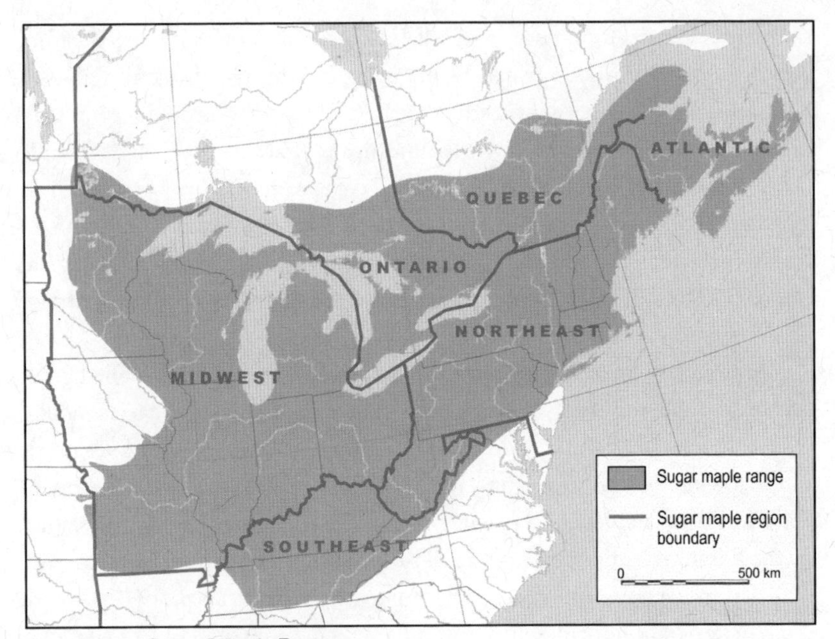

Figure 11.1. Sugar Maple Range.

Canada produces 80 percent of the world's maple syrup and related products and in 2013 produced 45.5 million litres of syrup. Within the country, the province of Ontario produced 2.04 million litres in 2013 with a value of $30.8 million (with Quebec being the clear leader at $346 million in 2013) (Agriculture and Agri-Food Canada 2014). While many NTFPs are largely unregulated, in Ontario maple syrup is increasingly regulated and is treated as an edible horticultural crop, with products offered for sale required to comply with Ontario Ministry of Agriculture and Food Regulation 119/11.[1] If producers wish to sell beyond provincial borders, they must also adhere to the Canada Agricultural Products Act through the Maple Products Regulations,[2] and producers are subject to inspection by the Canadian Food Inspection Agency. There are also unique labelling requirements for producers seeking different product certifications (e.g., Canadian Organic Standards, Forest Stewardship Council).

In Ontario, the production of maple syrup, like all forest-based activities, mirrors the province's dominant geologic, demographic, and land tenure patterns (see Bullock and Hanna 2012). These differences exemplify the management complexity facing broad-scaled NTFPs. The southern portion of Ontario has extensive stands of sugarbush situated on productive soils

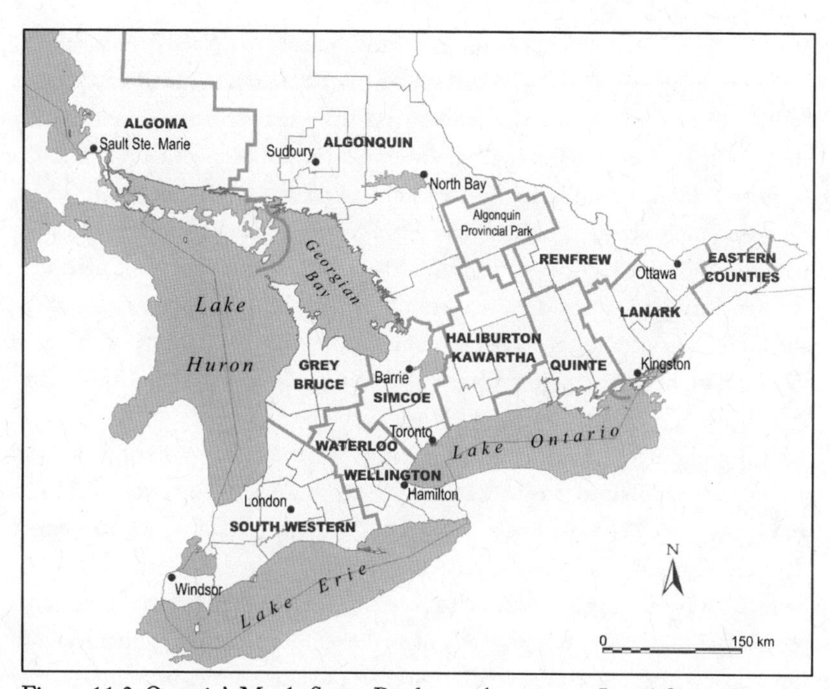

Figure 11.2. Ontario's Maple Syrup Producers Association Local Councils.

underlain by sedimentary rock, with the highest concentrations of producers located in the Waterloo-Wellington and Lanark County locals (Figure 11.2). While southern areas have significantly more access to nearby urban markets and 95 percent of the provincial population base, which is ever more willing to consume rural landscapes as tourists (Bullock and Hanna 2012), competition among producers is also more intense.

In the northern part of the sugar maple range, production is constrained by increasingly cold temperatures and the igneous rock and acidic soils of the Canadian Shield. Production is more limited to key pockets of good sugarbush, and production on Crown land is more common, as very little land in the north is privately owned. There are over 27 million hectares of Crown forest in the province, most of which is located in the northern part of Ontario (Harvey, Chapter 3). Where syrup production on Crown land can be undertaken under a long-term lease agreement, significant commercialization and local benefits can accrue. Without a formal agreement, sap harvesting often still occurs, but the more efficient (and costly) tubing sap collection systems cannot be set up, limiting the economic potential of the operation. Of the major syrup production regions in Canada, Ontario continues to have the least access to public lands (0.04 percent in 2011), compared to Quebec and New Brunswick (11 and 8 percent respectively) (OMSPA 2013: 3). Especially in northern Ontario, sugarbush on Crown land is always at risk from timber extraction activities, since most licensing agreements are oriented toward pulp and paper and dimensional lumber. Similarly, the predominant forest management practices in Ontario emphasize a neoliberal concern for the supply of timber consistent with a resource extraction model rather than the multiple uses associated with NTFPs. Low population densities, remote locations, limited access to markets, and high transportation costs can be significant barriers to building forest-based industries in northern regions (Harvey, Chapter 3; Palmer and Smith, Chapter 2; Murphy et al. 2012; Southcott 2006).

Throughout Ontario, Indigenous peoples have been gathering sap and producing syrup since pre-contact with European colonial states (Whitney and Upmeyer 2004; Goodman 2014). First Nations typically gather sap from designated reserve land where tree stands are managed communally by the band. First Nations peoples may also access trees from their broader traditional territory, if this land is not privately owned or under an active timber lease. In Ontario, we have had contact with a number of bands (ethics prevents us from revealing the locations) wherein members continue or are reclaiming maple sap and syrup practices. Land use is further complicated in First Nations contexts given that production may or may not be profit-oriented in nature. Where production occurs on reserve land, proceeds and profits are often (but not always) returned to the band for communal economic and cultural benefit rather than personal economic gain. Metis peoples do not have this communal land base and so tend to harvest on a combination of private and Crown lands across the province. Some Indigenous producers harvest sap and use it strictly for ceremonial purposes and may disapprove of commercial production. Others have wholly embraced rural/settler technological innovations, sometimes masking any Indigenous connection, while others blend the two traditions in a more overt way.

It's important to note that although maple syrup was initially an Indigenous technology shared with settlers, today their voice and preferences are relatively absent across the industry.[3] Widespread Indigenous producers are often isolated or have lost their Indigenous Knowledge (IK) due to colonial policies that outlawed many traditional practices and led to discrimination. Thus, harvesting sap and making maple syrup is connected to exercising Aboriginal and treaty rights and rectifying this historical legacy of cultural genocide (see also Casimirri and Kant, Chapter 4; Royal Commission on Aboriginal Peoples 1996, vol. 1: 250).

Development of the Indigenous and Rural/Settler NTFP Maple Value Models

Forest products and services have always been important to societies and today are recognized within the community forest movement as an important complement to timber extraction. NTFPs, including maple syrup, can contribute to subsistence needs and help diversify and supplement rural incomes. NTFPs include "the biological resources, products and services, other than timber, that can be harvested from forests for subsistence and/or trade" (Murphy et al. 2012: 43). NTFPs can be harvested from agro-forestry systems, primary and secondary forest, and forest plantations, and involve such products as "medicinal plants, fibres, resins, latex, oils, gums, fruits, nuts, foods, spices, flowers, crafts, dyes, construction materials, and fuel wood as well as related value-added products, tourism and festivals" (Murphy et al. 2012: 2; see also FAO 1995; Laird et al. 2010; Shanley et al. 2008). Further, for Indigenous peoples such as Canada's First Nations, Metis, and Inuit peoples, NTFPs have been foundational to their well-being and have been actively managed to provide food, clothing, and medicines, contributing significantly to their cultural and spiritual practices (Murphy et al. 2012).

NTFPs are often underrated as "minor" forest products with little value because they are frequently produced on a part-time basis in highly localized, hinterland locations and may be part of a hidden or subsistence economy. In Canada, maple products and other NTFPs such as berries, Christmas trees, honey, wild pelts, and mushrooms contribute up to $1.26 billion annually (Mitchell et al. 2010). However, at only 3 percent of the value of timber and pulp products, NTFPs are marginalized on the policy agenda. The predominance of the timber industry and the marginal status of rural and forest regions in highly urbanized western societies are also contributing factors (FAO 1995; Laird et al. 2010; Mitchell et al. 2010).

In contrast to NTFPs, in a primary industry such as forestry, a "staples" approach to economic value predominates and is focused on unfinished bulk commodities or raw products sold to export markets (Palmer and Smith, Chapter 2). A more comprehensive economic conceptualization of value, one that is useful in the NTFP context, was developed by Porter (1985), who defined a value *chain* as a firm's strategic activities that enhance profitability through the development of a differentiated product rather than a commodity, improved system efficiencies, and increased product quality. This approach incorporates a growing segment of today's consumers who are demanding products differentiated by their uniqueness (e.g., maple spa products), high quality (e.g., award-winning syrup) or third-party certification (e.g., Forest Stewardship Council/organic certification) (Agriculture and Food Council of Alberta 2004; OMSPA 2013; Shanley et al. 2008).

Across an industry, a value *system* links firms from raw product through intermediaries (channels) to the consumer in a stream of activities that promote quality and increase the product's value to the final consumer (Porter 1985). Value systems thrive through high levels of inter-firm communication and collaboration to achieve chain-wide goals (Agriculture and Food Council of Alberta 2004; Carpenter and Sanders 2009). These interactions build the trusting relationships and networks that underpin the development and enhancement of social capital (Inkpen 2005; Mohan and Mohan 2002).

On the rural/settler side, the value chain and system concepts can be expanded to develop a "maple value system" model that takes as its starting point the "triple bottom line" of sustainability. Consequently, our rural/settler maple value system model embraces a comprehensive definition of "forest values" that includes the "array of forest products, conditions, and human interactions with the forest that are deemed important by and for the community, whether for socio-cultural, economic, or ecological reasons" (Bullock et al. Chapter 1). As such, our maple value system model enables consideration of a much broader array of the social, economic, and environmental benefits produced via maple product production activities.

From an Indigenous perspective, we took a different approach that reimagines how we think about value systems. Based on stories and interviews collected to date, Indigenous values have been mapped according to an adapted Medicine Wheel model. The Medicine Wheel, also known as the Sacred Hoop, is a powerful symbol of Indigenous beliefs. As a conceptual tool, it is used and interpreted differently in many Indigenous communities. Its purpose is to respect the holistic approach that is characteristic of Indigenous belief systems and ways of knowing often referred to as Indigenous Knowledge (IK). The Medicine Wheel represents the cycles of life, interconnectedness of all things, and the harmony of the whole (Yearington 2010). In this work, we use it as an analytical tool to evaluate Indigenous values associated with maple trees and products and as a framework to represent these values.

Defining Community Values in Indigenous and Rural/Settler Syrup Production Contexts

In an effort to observe differential access to power and resources among groups of people involved in NTFP production, in this chapter we pay particular attention to the conceptualization of "community" and problematize narrow definitions of the concept that do not fully embrace the experiences of Indigenous and non-Indigenous people living in forest-based and agro-forest communities. The term "community" is often associated with a particular geographic space or territory such as a neighbourhood, municipality, or county. Indeed, this is a key type of community and one that is often assumed in the community forestry literature (Bullock and Hanna 2012). Another key type, a community of interest, develops around common practices or beliefs. These two types of communities often co-exist and overlap and are intertwined with other sets of social relations and factors operating at different scales. For instance, along the Indigenous–rural/settler continuum, settlers currently dominate the industry with their communities, maps, management preferences, and worldviews organizing space and production approaches. Yet, while the boundaries of Aboriginal treaty lands, traditional territories, and reserves as well as the cultural and political alignment of various First Nations and Metis peoples are less visible, increasingly these perspectives are complicating how maple syrup communities can be understood.

There are also other factors that add complexity to the concept of community. First, the literature on sustainability would suggest that attention to future generations broadens the definition of how a community is conceived. This requires communities to manage resources with consideration for long-term impacts and family heritage (Bullock and Hanna 2012). Second, regardless of the type of community, there are always within-group differences regarding values, perspectives, and access to resources and decision-making power. Attention must be paid to the richness and diversity that characterizes all communities. Third, western thinking tends to limit understandings of community to the human world. Bullock and Hanna (2012) hint at an expansion of this definition when they briefly refer to communities as socio-ecological systems.

In this work, we have adopted the phrase "all my relations" to reflect the different understandings of communities and social networks held by the Indigenous producers. The phrase "all my relations" is literally translated from the Lakota Sioux term *Mitakuye Oyasin* (Elk and Lyon 1990). It refers to the belief that all things are living sentient beings, and that we are all connected or "related." All my relations include not only the human family, but also plants, animals, and the forces of nature. As discussed in more detail below, this worldview was clearly expressed in our interviews with Indigenous producers and expands on the view that socio-ecological systems can include both humans and trees. And it is interesting to note that scientists are beginning to catch up with the traditional belief in the social lives of plants (Biedrzicki 2010; Dudley and File 2007).

Based on seven years of research, especially a recent round of thirty-four interviews, in the remaining sections of this chapter we outline how maple syrup as an NTFP is valued and understood in rural/settler and Indigenous communities in Ontario, Canada. After providing an overview of the research trajectory and methods, we outline the results and the two maple syrup value models that were developed for this project. In the final section, we draw insights across these two models and offer suggestions for community forest operations that are considering moving into NTFPs as well as value-added products and services.

Research Approach and Methods

The research presented here emerges from several related projects focused on exploring the ways in which maple syrup contributes to sustainability and resilience in rural, agro-forest, and forest-based communities and the potential impacts of climate change. Through our program of research we have worked closely with stakeholder groups including maple industry associations, settler and Indigenous producers, and government officials. From the outset, we adopted a transdisciplinary methodology that incorporates multiple ways of knowing. We draw from multiple academic disciplines across the humanities, social sciences, and physical sciences as well as a range of stakeholder perspectives (Brown et al. 2015; Murphy et al. 2009, 2012; Murphy 2011).

For this sub-project, we set out to undertake a sector profile of the maple syrup industry, identifying the value system's key players, processes, and inter-intra relationships to provide baseline data on the current status of the Ontario industry. We undertook thirty-four semi-structured interviews with participants from across Ontario's maple production regions, representing different sizes and types of operations. Interviews were divided into three segments. Fifteen interviews were conducted with rural/settler producers (Morin), fifteen interviews with Indigenous producers (interviewed by Smits, analyzed and written up by Chretien) and four with other key informants (Murphy). Interviews were digitally recorded and later transcribed and analyzed in the NVivo qualitative data analysis program. Data was coded deductively using predetermined themes, with inductive coding added, where warranted, as the analysis unfolded.

The program of work is overseen by an advisory committee who comment and contribute to all aspects of the research. At an inaugural advisory board meeting, the original concept for the Indigenous maple values model was suggested by Elder Charles Restoule (Dokis First Nation), and the group put forward initial ideas about the rural/settler value system model. At a later date, David Chapeskie (International Maple Syrup Institute) provided significant input into the latter model. In these ways our research follows a participatory, community-based approach.

Value System Models

Below we present the two maple system models developed through several iterations during the research process. We then provide supporting data to help contextualize and explain the models. The two models were intentionally developed separately to allow different knowledges and ways of knowing to inform their orientation. It was particularly important to highlight distinct Indigenous understandings of maple since these voices have receded to the background as the settler community has come to dominate the maple syrup industry.

Indigenous Maple Syrup Value Model

As mentioned above, the model developed from Indigenous perspectives is based on the Medicine Wheel, which emphasized the cyclical nature of maple practices and how these could be best understood (Figure 11.3). The Medicine Wheel served as a research tool, an analytical approach, and a framework to represent the data collected in our interviews. Elder Charles Restoule's ideas of how to represent maple values using the Medicine Wheel were echoed in our interviews with Indigenous producers. The cycle of life as understood through maple practices was mentioned over and over, leading us to further develop the model to reflect these stories and beliefs. As Elder Restoule suggested, the model starts in the East, reflecting Indigenous beliefs for beginnings, and progresses through the yearly cycle of maple production by moving around the adapted Medicine Wheel. The model was an effective way to map the values that were shared by our interviewees because it emphasizes process, relationships, and spiritual beliefs. The centre represents Indigenous values as expressed through the concept of "all my relations" as defined above.

In keeping with the belief of the renewal of life, the sap, which is called "sweetwater," is used medicinally, especially for pregnant women, who are considered the givers of life. The use of sweetwater as medicine and as a cleansing agent is widespread in Indigenous communities. As mentioned earlier, maple sap has recently entered the health drink market much in the same way as coconut water, and with similar uses. However, this practice is frowned upon by Indigenous traditionalists, who insist that sweetwater is medicine and should never be sold, or worse, widely marketed. It is considered sacred, not a commodity.

The opening of maple syrup season in Indigenous contexts carries with it very distinctive and diverse beliefs and ceremonies, some of which are deeply rooted in traditions that can go back hundreds of years. For some producers, preparation for the maple season actually begins with winter ceremonies praying for "new babies" in the spring. These ceremonies are intended to meet the responsibilities of an ongoing relationship between people, trees, and the environment. Since maple syrup is the first product to be harvested in the spring, it represents new life, and for some Indigenous people, the beginning of the new year. For example, in Aniishnaabe belief systems, the calendar consists of thirteen "moons" rather than twelve months. March is the "sugar moon," *Zhiishbak Geezis*, which marks the beginning of a new annual cycle.

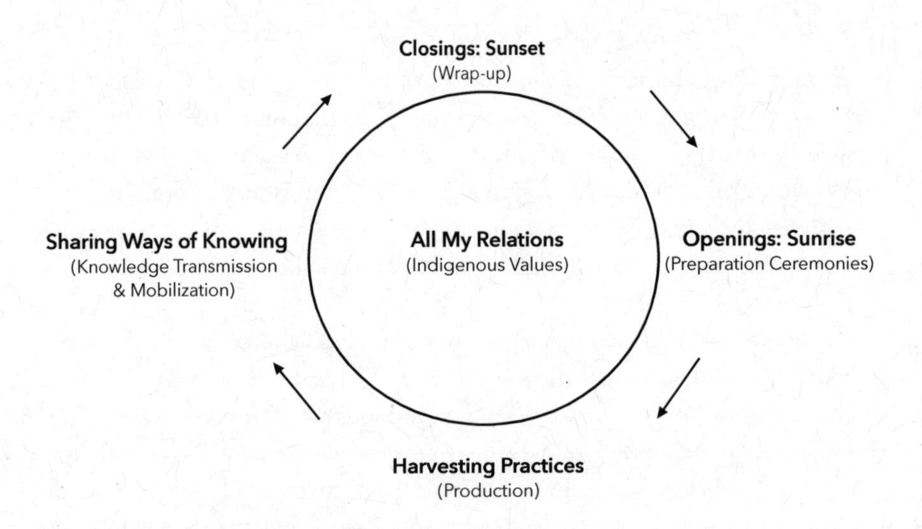

Figure 11.3. Indigenous Perspectives of the Maple Value System.

Opening ceremonies in other Indigenous communities may differ in practice but were very similar in terms of beliefs, as evidenced by the terms used for these ceremonies, such as "waking up the earth." Spring, ceremonies, spiritual beliefs, and the new year were all terms that were repeated by the Indigenous producers we interviewed. Maple sap harvesting and syrup making as a cultural practice serves as a marker for the cycle of death and life.

It is important to note here that not all Indigenous producers we interviewed followed traditional practices or would even be seen as Indigenous producers in the eyes of the public. Some producers preferred not to follow traditional ways. By contrast, some wished they could follow more traditional practices but the IK had been lost throughout the generations. One producer mentioned a "spiritual renaissance in Indian country," alluding to the fact that many are currently reclaiming lost cultural practices and the IK that accompanies them. There are many reasons for the loss of maple practices in Indigenous communities and their relative invisibility in today's industry, most of which can be connected to the legacy of colonialism. The intimate link between government policies and Indigenous cultural practices of all sorts, including making maple syrup, is of particular relevance to this chapter.

Many interviewees indicated that there were periods when making syrup was not practised and that they are currently reclaiming knowledge from the elders who recall it from previous generations. One interviewee told a story about being forbidden from having ceremonies by the Indian agents and that longhouses were locked to prevent them from gathering for the maple syrup season. This is why knowledge was not passed down openly. Some people did continue to harvest sap but kept everything hidden to keep the men out of jail. Another producer recounted that the Jesuit missionaries also prohibited harvesting maple sap and making syrup because it kept the people in the bush and prevented them from going to church. Colonial mechanisms aimed at assimilating Indigenous peoples deeply affected Indigenous maple cultural practices. This partially explains the discrepancies in the type and depth of IK held by today's Indigenous producers.

Further, as IK is reclaimed in a contemporary world, it is reinterpreted and reconstructed to some extent, resulting in a blending of old and new beliefs, traditions, and practices. Thus, the harvesting practices of Indigenous producers are quite diverse and demonstrate a mix of traditional and modern practices. For example, one large commercial producer uses lines, tubing, and a vacuum system but keeps a few buckets and spigots to monitor the harvest. He uses these trees to assess how the trees are doing. Since the lines are a closed system, there is no way of knowing when to stop tapping. In his opinion, it runs the risk of drying out the trees. When the sap gets milky in the buckets, he knows it is time to stop harvesting.

Other producers felt that tubing and vacuum systems are harmful to the trees and they refuse to use this technology. Some traditionalists also felt that this type of harvesting diminished the sacred nature of maple sap and syrup and its medicinal properties. The use of tubes and vacuums are seen as taking from the trees rather than the trees giving their sap willingly. A deep concern for the trees was expressed. Traditional harvesting practices revealed a close relationship that included getting to "know" the trees; in some cases, individually. This intimate relationship is further evidenced by the notion that the trees are the ones who taught humans how to make the syrup and who present people with their gift of sap.

The profound differences between rural/settler and Indigenous producers were most clearly revealed in the ideas of sharing knowledge and community. For example, one producer recounted that the ceremonies are a communication with the entire earth, the air, water, and humans; in other words "all my relations." This is how humans build their relationships with the trees and the environment. The trees are not only considered to be "social beings" but they are part of the human social network at the family level. The trees are even considered to have families of their own. One producer referred to them as having "uncles and aunties" and even their own nations. In sharing their sap, the trees were visiting and teaching their human relatives.

The intimate relationship that some traditional Indigenous producers have with maple is connected to the idea that the sap is medicine, not a commodity. This belief influences every aspect of traditional production and prevents many producers from using more recent technologies and production methods as well as selling their sap or syrup. In this context, maple products are for medicinal, ceremonial, personal, and community uses only. Any change of practice is seen as tainting the product and taking away its sacred and medicinal values.

By contrast, other Indigenous producers do not necessarily hold the same views and are quite comfortable selling maple syrup commercially. They still maintain a respectful relationship with the trees and environment; however, they also feel that maple production is a sustainable activity that can bring much-needed economic activity, especially in isolated communities. To do so, they use the technology that best aligns with their IK and worldviews. Commercial maple production in Indigenous communities is complicated by the communal nature of land ownership on reserve. In some cases, individual members of the band make and manage the syrup operation on behalf of the band, share some of the final product, but retain some of the financial profits for their own family. Indigenous producers who are not part of a band or on a reserve (e.g., Metis producers) may share with their families and communities in a more informal way.

In some communities, where the commercial operations are owned by the band, all economic proceeds are returned to the band for communal use, with some of the maple products also shared out with members. These enterprises blend traditional subsistence systems with a market-based model. This is not to say that all Indigenous producers operate in these ways or hold the same beliefs. In an economic sense, producers' attitudes ranged from the very traditional view of never selling any maple sap, syrup, or other related products, to a combined approach of selling a little to support the activity, through to undertaking a fully commercialized venture viewed as an economically sustainable business.

In terms of closing the season, most producers follow typical cleanup and sugarbush management processes. Commercial producers clean their tubing systems and equipment and store them for the following year. The bush is examined for diseased trees, which may or may not be culled. Smaller backyard producers will also survey their bush. One producer commented that cleaning the small brush around trees helps the trees grow much bigger. Having said that, for those who follow a more cyclical understanding of the harvest, cleanup does not necessarily mark the end of the season, since the cycle is ongoing. One producer likened the cycle to the chicken-or-egg dilemma. Beginnings and endings are not clearly defined in a circular model.

On a final note, the idea of "all my relations" is clearly lived and guides the values that are associated with maple trees and their gift to us as humans. Economic factors were far from the main value expressed by most producers. The reclaiming of culture, history, identity, and medicinal and spiritual values far outweighed economic gain as the major factor for maple practices. Many producers see making maple syrup as a way of reclaiming their Indigenous identity, history, and culture, and of re-establishing a relationship with their ancestors.

Rural/Settler Maple Values System Model

The rural/settler model builds from ideas around Porter's (1985) value system to incorporate ideas associated with sustainability. The model is interconnected and iterative, centred on economic, social, cultural, and environmental core values, and contextualized by dominant trends in these realms and by

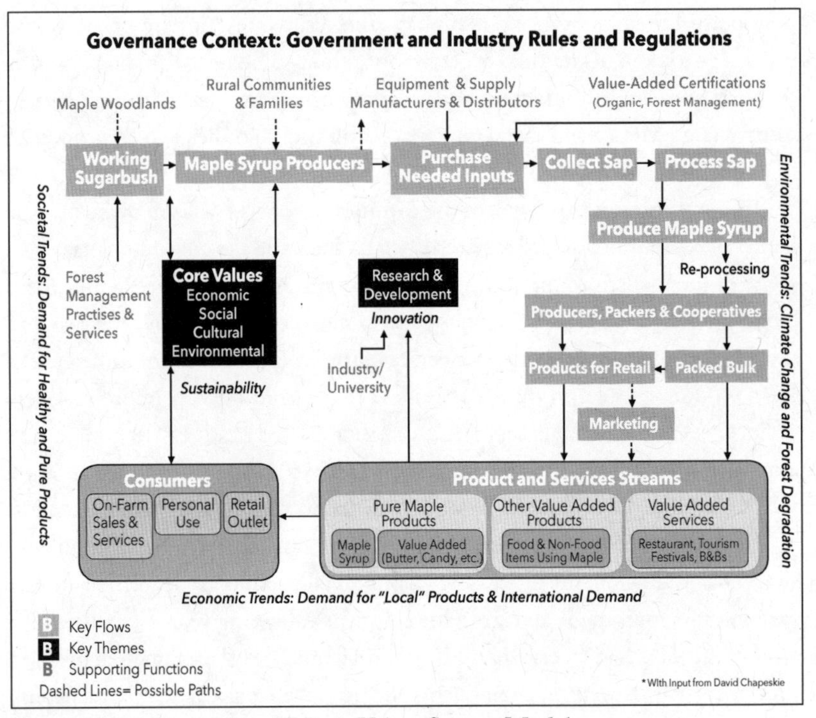

Figure 11.4. Industry-based Maple Values System Model.

the governance framework within which maple is embedded. It starts with the maple woodlands and then moves through communities, supply channels, production, final products, and consumers. Innovation, denoted by the research and development box, is undertaken by a range of actors throughout the value system (Figure 11.4).

In terms of the environment, Ray Bonenberg, past president of the Ontario Maple Syrup Producers Association (OMSPA), is very clear that the forests are the "goose that lays the golden egg" (Goodman 2014). He advocates for prudent forest management practices to manage the resource for the long term and has undertaken Forest Stewardship Council certification in his own operation. In addition, the Ontario industry has developed a set of sugarbush best management practices. Interview participants spoke of their pleasure when getting out on the land in the early spring and about their connection to these forest environments. Producers develop an intimate relationship with their trees and talked about adjusting tapping and management approaches to reflect any noted stressors. Multiple uses are often provided by larger operators, who may offer trail systems and other outdoor activities, such as hay rides, to visitors.

Respondents maintained that significant challenges facing the sugarbush are logging pressures, invasive species, forest degradation, and climate change. One example of the predominance of timber-oriented perspectives is that producers find it difficult to locate foresters who know how to manage forests for uses other than timber. For instance, one respondent suggested that issues less typically addressed by forest professionals are managing for multiple forest uses and pruning trees to maximize sugar maple sap sweetness by encouraging tree crown spread (rather than pruning trees for straight trunks harvestable for lumber). Again with reference to the dominance of timber, one interviewee from the north commented that sugaring operations could expand more rapidly if access to Crown land through long-term leases could be more easily obtained.

As generally understood by industry members, the production community displays many subgroups and interests and can be seen as existing along a continuum. At one end of the scale, for small and backyard producers (fewer than 1,000 taps), maple was noted as providing a product that can be used for subsistence and shared with family and friends. Buckets and smaller or simpler evaporators are typically used. At the midpoint are producers who have larger operations (about 2,000–5,000 taps) and significant investment in equipment and tubing systems. At this level, production is usually a sideline operation undertaken along with other farming or economic pursuits and typically focused on the spring production season. Maple provides a welcome cash infusion in the early spring, and some value-added pure maple products may be produced. At the other end of the scale, for the largest operations (10,000–20,000 taps), maple is often undertaken as a year-round business. In these operations very substantial capital investments are made, operations can provide significant local employment, and producers may undertake extensive research and development to bring new products to market (e.g., some have their own government-inspected test kitchens). Value-added often extends into a range of innovative maple-enhanced products and services, including foodstuffs, pancake houses, bed and breakfast operations, and so on. Across the continuum, beyond the substantial costs of equipment and land, most respondents are concerned about the costs of fuel, labour, and supplies (e.g., packaging). In Ontario, there is also a burgeoning group involved in the industry that includes individuals who may or may not be producers themselves and instead primarily deal in bulk syrup and act as packers and exporters, or focus on artisanal gift products. The packers act as an important conduit to the final consumer for those producers who enjoy making syrup but have no interest in dealing with the public or undertaking marketing activities.

Participants were very clear that while maple can provide a good income stream it goes far beyond this by contributing to a valued way of life, relationships, and forest-based lifestyle. For instance, when thinking about future generations, one participant, a producer with a larger, year-round business, clearly expressed the connection between the financial aspects of the business and long-term sustainability, noting that they were in it to produce for the next 100 years, so it was important to look beyond yearly economic gains to protect the health of trees. This interviewee suggested that the social value of family, community, and staff were more important than the economics. This also meant taking "expert" advice with a grain of salt because experts don't necessarily know as much about particular local sugarbushes or maple practices as the property owners do. We heard similar views expressed repeatedly throughout the seven years of our work.

Most of the producers' operations are family businesses, and family networks are a key component of the maple syrup value system. Needed labour is often volunteered by family members, and extended family networks help in marketing the syrup. Needless to say, family heritage is an ongoing conversation in the maple community. Some respondents were concerned that they are getting older and that there might not be someone interested in learning about maple and inheriting the business. When "new blood" take over an existing business (or start up a new operation), other producers are relieved knowing that the operations will continue and/or expand. New entrants are frequently celebrated and supported by the whole maple community. In other cases, some producers traced their heritage back to settlers from the 1800s and were actively grooming the next generation to take the reins.

Supporting the farm-level activities are equipment and supply manufacturers and dealers. As explained by respondents and noted by the research team on multiple occasions, equipment dealers are a key part of the close-knit maple value system and regularly share their expertise with various others involved in the maple syrup industry through one-on-one visits and trade show presentations. As Ontario is experiencing significant growth, the large equipment manufacturers from Quebec and the United States are actively soliciting business in the province and setting up local dealers. Equipment manufacturers undertake extensive research and development. They have developed ever more tree-safe and efficient tapping systems as well as evaporators and related equipment that have significantly reduced the labour commitment and energy costs of maple operations. For example, buckets need to be inspected regularly to ensure that they have not overflowed; by switching to a tubing system, the sap is collected at a central location and kept in a large holding tank. The innovation and services provided by dealers differentiate their products as well as contributing to innovation and robustness across the whole system. Useful innovations tend to trickle down through the industry, leading to system-wide changes over the long term.

For producers, product differentiation can occur in two key ways: (1) across a sector in relation to substitutable products and (2) within an industry between competing firms. Maple products are part of the larger sweetener sector that includes cane sugar, honey, agave syrup, and so on. The maple industry has been undertaking research related to the health benefits of maple syrup (e.g., antioxidants and trace minerals) and has developed materials that explain these benefits (see IMSI 2012). Respondents also noted that there is no point in trying to compete with very inexpensive substitutes such as cane sugar. Instead, maple syrup should be thought of as high-value product like good olive oil, and the contribution of maple to the local economy should be highlighted. For instance, the idea that the taste of maple is affected by "terroir" (the influence of a location's geology, soils, and climate on a crop) and the craftsmanship of the producer is gaining traction in the industry and is providing mechanisms to differentiate maple from other sweeteners as well as to differentiate between the syrups from different producers.

Within the industry, other forms of differentiation can include third party certification and specialization. The value system literature suggests that when effort is invested in a way that is hard to duplicate, the firm can gain competitive advantage. Maple industry members are undertaking differentiation activities such as research and development to create maple-enhanced products (e.g., spa product) or to develop specialized information (e.g., overseas market connections). However, interviewees did not tend to articulate such activities in terms of competition. Most viewed other producers as part of their larger community rather than competitors and suggested that there was plenty of room for everyone who has an interest in maple. That said, some producers in southern Ontario, where there are higher densities of producers, did note some minor tensions or friendly competition among neighbouring operations.

As noted by many respondents, third party certification such as FSC woodlot or organic certification is still quite uncommon. Although this does downplay the guidelines demanded through third party certification, many asserted that maple is a pure and natural product derived from trees and that forest stewardship is common, so the need for outside verification of their practices was redundant and costly and many farmers balked at having to undertake this type of bureaucratic activity. Those who undertook certification did so to access particular markets such as health food stores and corporate sales or to differentiate their product, especially for online or overseas sales. These individuals maintained that certification did not necessarily increase the price they could charge, but opened up new markets.

Another type of third party verification is becoming a facility approved by the Canadian Food Inspection Agency. This is seen as an arduous process usually only undertaken by those who want to export outside of the province or who feel that this outside verification boosts buyer confidence and increases sales. Interestingly, a recurring idea expressed by interviewees was that membership in OMSPA was itself considered a form of product differentiation that could be proudly displayed on labels, since the organization was said to promote high-quality production and food safety standards.

Regarding information needs, two dominant trends were noted by respondents. First, the need for marketing strategies related to both the domestic and international markets was a recurring theme in the interviews. Having current information to help expand into new markets, especially in other countries such as Japan (stated to be one of the world's largest importers of maple syrup) was an issue raised by some interviewees. Many respondents also noted that the Ontario industry has enormous potential. Participants estimated that less than 5 percent of the sugar maple trees in Ontario are tapped and stated that a good percentage of the Ontario market is supplied by Quebec syrup. Further, respondents directly connected maple's potential to growing environmental discourses, explaining that maple is renewable, "green," and doesn't harm the land. Yet, despite this potential and the increasing levels of production, selling the product to consumers was a struggle for some, especially in areas of southern Ontario where competition was higher or in more northern areas where remoteness hindered market access.

Word-of-mouth advertising, simple signage, and direct contact with the consumer are dominant forms of marketing mentioned by respondents. While these approaches are working well for many, there is also a growing trend, especially among the younger producers, to undertake web-based and social media marketing. And, although today's large commercial operations look nothing like the iconic "sugar shacks" of old, nods to this history are still displayed on much of the marketing material, playing into marketing and buyer demand for consumable heritage.

In addition, producers strongly believe that one poor batch of syrup or a media-hyped maple-related food safety scare could easily undercut decades of relationship building. Trust in the family name and quality guarantees are considered key to successfully selling their maple products. Respondents were especially concerned that producers not under the OMSPA umbrella might be less knowledgeable about best practices and more prone to having a food safety issue such as mould. Yet the public would not likely distinguish between member and non-member producers if a scare were to occur, and this could damage the whole industry. OMSPA is well aware of these concerns and is undertaking concerted efforts to expand its membership and develop effective marketing strategies for producers within the province.

As a direct result of the ongoing technological advancement, respondents suggested that there is a concurrent need for advanced information and on-the-ground support for using and maintaining the equipment. Attending various industry events and workshops, and informal information exchanges between neighbouring operations, are considered crucial to running a professional operation. Ongoing communication promotes a tightly knit producer community within the maple syrup industry. Conversely, respondents also suggested that Ontario dealers, with head offices located in other jurisdictions, typically have very extensive catchment areas. This could lead to some difficulties in accessing information, parts, and service.

Discussion and Concluding Remarks

We have discussed maple syrup as an NFTP and presented two models that map rural/settler versus Indigenous maple value systems. Developing these models separately has been viewed quizzically by some rural/settler producers, including individuals from our advisory committee. Indigenous producers, in contrast, were adamant that we take this approach so that their distinctive perspectives could be documented. This seemed particularly important considering the current dominance of the rural/settler producers. We fully admit that there are overlaps and gaps and that we present only a partial picture. Yet we have found that the process of trying to find some fuzzy boundary between the rural/settler and Indigenous understandings of the maple resource has been useful in highlighting the broad differences in how worldviews and ways of knowing shape understandings of the maple resource as well as drawing attention to the impact of historical legacies. There is no doubt that if successful community forest initiatives wish to include Indigenous partners, their distinct values and beliefs would need to be seriously considered and included in this approach.

Although we have presented these models as two distinct ways of knowing to demonstrate the differences, overlaps are clearly evident. Both sets of interview participants displayed strong forest stewardship principles and believed that the value of maple extended beyond economics into the realms of social and cultural sustainability. Indigenous producers involved in commercial operations straddle both of these models, and we could imagine a third model bringing these two mappings together.

A comprehensive analysis of the entire value system underpinned by the range of identified values is a useful way to envisage a given NTFP sector and its potential. It is interesting to document that although all producers start out with the same resource—maple sap—the availability of that resource, and how it is understood and used, varies substantially across producers. Community forestry initiatives seeking to incorporate NTFP harvesting can look to the details of this case study to understand some of the challenges and opportunities they might need to address. For instance, our analysis revealed that for some of the Indigenous producers, the spiritual/cultural and ceremonial value precluded commercial uses of maple sap, including market production.

Ongoing challenges facing the commercial maple industry include: further developing markets relative to other sweeteners; the need for better marketing strategies; gaining access to productive sugarbush stands on Crown land; and having policy needs addressed by various government bodies. On this latter front, the industry has been working for several years to instigate harmonized maple grading standards and language across Canadian and American political jurisdictions to reduce consumer confusion about types and colours of maple syrup. As an NFTP, maple is not typically a policy priority; thus it has been an arduous task to resolve this issue. For other sectors, institutionalizing key policies that enable and support the NTFP may also be needed.

Consideration for the various permutations of geographic and interest-based communities will need to be assessed by community forestry projects interested in NTFPs; a focus on the former and a lack of attention to diversity could leave simmering below the surface a host of differences that may undermine the NTFP sector. Ontario producers can be divided along multiple axes, including Indigenous vs. rural/settler; OMSPA vs. non-OMSPA members; north vs. south; small/medium/large operations; Quebec vs. everyone else; tapping on Crown land vs. private land; and backyard/subsistence/ceremonial vs. commercial.

Maple syrup demonstrates the potential that NTFPs can contribute to the triple bottom line of sustainability in Indigenous, rural, and forest-based communities. For rural/settler producers, ideas about community and sustainability are intertwined and extend to future generations, since this NTFP requires management of the sugarbush for the long term and ownership succession is a key concern. With Indigenous producers, ideas about community are pushed still further to include the concept of trees as sentient beings having their own families who share their sap with the human community.

We developed the Indigenous and rural/settler maple value models with substantial and long-term input from our transdisciplinary team and interview respondents—we could not have imagined these mappings on our own. To that end, we recommend that if community forest organizations are considering developing an NTFP, broad consultation will be crucial to uncover and document the range of values held by interested groups.

We suggest that attention to the following questions could be useful during such consultations:

(1) How is the NTFP understood, processed, and ultimately used, and how does this vary across harvesters?
(2) Who are the harvesters of a NTFP, what are their goals and aspirations, and do differences exist related to underlying worldviews and ways of knowing?
(3) Where is the resource harvested and how does this vary across space and production methods?
(4) Who are the ultimate users of the resource and how do they obtain the product?
(5) What are the substantive contextual issues (e.g., tenure, previous forest degradation, power structure, access, governance, technological, marketing) that enable or constrain sustainable harvesting and use?
(6) What are the historical socio-cultural patterns or environmental legacies that structure and influence current realities?
(7) What are the short-term and long-term socio-economic and political trends and public discourses that could impact the resource's future potential?
(8) Can resource harvesting contribute to multiple values and uses that are sustainable over the long term?

Answering these and other emerging questions through inclusive consultation processes will help identify and resolve both normative and practical issues associated with localized NTFP production and consumption.

Concerted effort to include key local knowledge holders and ways of knowing is critical to developing a nuanced understanding of what is at stake for NTFP harvesters, identifying potential points of conflict, and providing a forum for marginalized voices (Casimirri and Kant, Chapter 4). In the case of maple, we are using our findings to sensitize rural/settler members to other voices and ways of knowing and we are helping to connect isolated Indigenous producers. We are also producing materials to support those communities who wish to reclaim or enhance their sap and syrup production practices. Something as simple as acknowledging traditional territory at industry events begins the process of developing a fruitful relationship with Indigenous communities.

Government policies in the past impacted Indigenous maple production in profound ways both economically and culturally. However, the extent to which Indigenous maple culture and economy were affected is not well understood. Some interesting questions remain to be addressed. Early sources note that maple sugar, not maple syrup, was the preferred commodity in the nineteenth century before processed sugars became widely available later (Spencer 1913; Butterfield 1958; Schuette and Idhe 1946). There is some preliminary evidence that suggests large-scale Indigenous maple sugar production was an important commercial industry in the nineteenth century. For example, historic information about maple sugar production on Manitoulin Island, which was mostly inhabited by Indigenous people at the time, presents surprising numbers. Cadieux and Toupin (2007: 9) cite maple sugar production on Manitoulin Island in 1846 at 86,000 pounds. In a report from the same era, Fortin (1865: 152) stated that the island produced 500,000 pounds annually. *Farm and Mechanic* reported that Manitoulin Island exported over 100 tons of maple sugar in an unfavourable season (Maple Sugar 1848: 41). By contrast, the role of Indigenous producers in today's Ontario maple syrup industry is negligible. There is no doubt that maple practices continue in many Indigenous communities, but producers are not visibly present in the public market. Whether or not this will change remains to be seen. One question that needs to be answered is whether or not government policies were the main factor that prevented a thriving Indigenous maple industry from continuing into the twentieth century and beyond. Answering this question would go a long way toward creating conditions favourable for community forest NTFP production in Canada today.

Acknowledgements

We gratefully acknowledge the financial support of Wilfrid Laurier University, the Ontario Ministry of Agriculture, Food and Rural Affairs, the Social Science and Humanities Research Council of Canada, and the Ontario Maple Syrup Producers Association. We also wish to thank all the individuals who gave their time so generously for this project as well as the ongoing support from our advisory committee.

Notes

1 Ontario maple regulations: http://www.omafra.gov.on.ca/english/food/inspection/maple/othr-mple-lbl-reg11911.htm.

2 Federal maple regulations: http://laws-lois.justice.gc.ca/eng/regulations/C.R.C.,_c._289/.

3 Although Awazibi Maple Syrup is a notable exception: http://kzadmin.com/Awazibi.aspx.

References

Agriculture and Agri-Food Canada. 2014. "Statistical Overview of the Canadian Maple Industry 2013." Accessed 11 January 2015. http://publications.gc.ca/collections/collection_2016/aac-aafc/A71-40-2013-eng.pdf.

Agriculture and Food Council of Alberta. 2004. "Value Chain Guidebook: A Process for Value Chain Development." Accessed 29 August 2014. http://www1.agric.gov.ab.ca/$department/deptdocs.nsf/all/agp7974/$FILE/valuechain.pdf.

Biedrzycki, M.L. 2010. "Root Exudates Mediate Kin Recognition in Plants." *Community Integrated Biology* 3: 28–35.

Brown, L.J., D. Lamhonwah, and B.L. Murphy. 2015. "Projecting a Spatial Shift of Ontario's Sugar Maple Habitat in Response to Climate Change: A GIS Approach." *Canadian Geographer* 59 (3): 369–81.

Bullock, R.C.L., and K. Hanna. 2012. *Community Forestry: Local Values, Conflict and Forest Governance.* Cambridge: Cambridge University Press.

Butterfield, R.L. 1958. "The Great Days of Maple Sugar." *New York History* 39 (2): 151–64.

Cadieux, L., and R. Toupin. Eds. 2007. *Letters from Manitoulin Island: 1853–1870.* Translated by S. Pearen and W. Lonc. Ottawa: William Lonc.

Carpenter, M., and G. Sanders. 2009. *Strategic Management: A Dynamic Perspective Concepts and Cases.* New Jersey: Pearson Education.

Dudley, S.A., and A.L. File. 2007. "Kin Recognition in an Annual Plant." *Biology Letters* 3: 435–38.

Elk, B., and W.S. Lyon. 1990. *Black Elk: The Sacred Ways of a Lakota.* New York: Harper Collins.

[FAO] Food and Agriculture Organization of the United Nations. 1995. *Non-Wood Forest Products for Rural Income and Sustainable Forestry.* Non-Wood Food Products Series #7. Rome: United Nations.

Fortin, P. 1865. *Sessional Papers. Volume II. Fourth Session of the Eighth Parliament of the Province of Canada.* Ottawa: Hunter Rose and Co.

Goodman, A., Director. 2014. "Progress and Potential: Ontario's Maple Industry." Accessed 17 January 2017. https://www.youtube.com/watch?v=zLjyvmweI8Q.

[IMSI] International Maple Syrup Institute. 2012. "Nutrition and Health Benefits of Pure Maple Syrup." Accessed 2 September 2014. http://www.internationalmaplesyrupinstitute.com/uploads/7/0/9/2/7092109/__nutrition_and_health_benefits_of_pure_maple_syrup.pdf.

Inkpen, A. 2005. "Social Capital, Networks and Knowledge Transfer." *Academy of Management Review* 30 (1): 146–65.

Laird, S.A., R.J. McLain, and R.P. Wynberg. 2010. Introduction. *Wild Product Governance: Finding Policies that Work for Non-Timber Forest Products,* 1–14. London: Earthscan.

McGregor, V. 2013. Project Advisory Board. Personal communication, September.

Maple Sugar. 1848. *The Farm and Mechanic* 1 (2): 41.

Mitchell, D.A., S. Tedder, T. Brigham, W. Cocksedge, and T. Hobby. 2010. "Policy Gaps and Invisible Elbows: NTFPs in British Columbia." In *Wild Product Governance: Finding Policies that Work for Non-Timber Forest Products*, edited by S.A. Laird, R.J. McLain, and R.P. Wynberg, 113–34. London: Earthscan.

Mohan, G., and J. Mohan. 2002. "Placing Social Capital." *Progress in Human Geography* 26: 191–210.

Murphy, B.L. 2011. "From Inter-Disciplinary to Inter-Epistemological Approaches: Confronting the Challenges of Integrated Climate Change Research." *Canadian Geographer* 55 (4): 490–509.

Murphy, B.L., A. Chretien, and L.J. Brown. 2009. "How Do We Come to Know? Exploring Maple Syrup Production and Climate Change in Near North Ontario." *Environments* 37 (1): 1–33.

———. 2012. "Non-Timber Forest Products, Maple Syrup and Climate Change." *Journal of Rural and Community Development* 7 (3): 42–64.

[OMSPA] Ontario Maple Syrup Producers Association. 2013. "The Economics of Maple Syrup Production in Ontario." Accessed 1 September 2014. http://www.ontariomaple.com/images/pdf/%C9coR_OMSPA_Econ_Imp_Final.pdf.

Porter, M.E. 1985. *Competitive Advantage: Creating and Sustaining Superior Performance.* New York: Free Press.

Royal Commission on Aboriginal Peoples. 1996. "Chapter 9: The Indian Act." In *Report of the Royal Commission on Aboriginal Peoples: Looking Forward, Looking Back.* Vol. 5, 235–308. Ottawa: the Royal Commission on Aboriginal Peoples.

Schuette, H.A., and A.J. Idhe. 1946. *Maple Sugar: A Bibliography of Early Records.* Madison: University of Wisconsin.

Shanley, P., A. Pierce, S.A. Laird, and D. Robinson. 2008. *Beyond Timber: Certification and Management of Non-Timber Products.* Bogor: Harapan Prima Indonesia.

Southcott, C. 2006. *The North in Numbers: A Demographic Analysis of Social and Economic Change in Northern Ontario.* Northern and Regional Studies Series #15. Thunder Bay, Ontario: Centre for Northern Studies, Lakehead University.

Spencer, J.B. 1913. *The Maple Sugar Industry.* Dominion of Canada Department of Agriculture.

Teitelbaum, S., and R. Bullock. 2012. "Are Community Forestry Principles at Work in Ontario's County, Municipal, and Conservation Authority Forests?" *Forestry Chronicle* 88 (6): 697–707.

Whitney, G., and M. Upmeyer. 2004. "Sweet Trees, Sour Circumstances: The Long Search for Sustainability in the North American Maple Products Industry." *Forest Ecology and Management* 200: 313–33.

Yearington, T. 2010. *That Native Thing: Exploring the Medicine Wheel.* Nepean, ON: Borealis Press.

The Economic Advantage of Community Forestry

David Robinson

Although community forestry is generally thought to be more equitable, environmentally sounder (Bray et al. 2006), and even "a solution to the vast and ongoing process of deforestation" (Schusser 2012: 214), it is rarely presented as a serious economic challenge to large-scale industrial forestry. Economic theory suggests, however, that community forestry is economically superior to industrial forestry and will deliver benefits that industrial forestry cannot.

As a business model, community forestry is expanding around the world (Arnolds 2001; Devkota 2010). It has been developing most rapidly under pressure from Indigenous populations. It is, however, a relatively minor though growing phenomenon in Canada. The predominant model for Canadian forestry is industrial production by private firms on Crown land. Of Canada's $30.4 billion in forest product and building and packaging exports in 2011, virtually none came from community forestry (Statistics Canada 2015).

Not only is the institutional form underdeveloped, the theoretical machinery for thinking about economics of community forestry is almost non-existent. Community forestry will involve new contractual arrangements; new interpretations of individual and collective rights with respect to local forests; new marketing arrangements; specialized education for management; new opportunities for foresters to experiment with silvicultural techniques (see, e.g., Leslie, Chapter 10) and intensive local forestry; innovation in labour markets; new techniques for collaboration and democratic planning; and new applications of economic theory to the new and evolving institutions.

This chapter considers community forestry from an economic point of view, laying out a number of conventional economic arguments that lead to the conclusion that community forestry is likely to be more productive and more efficient than the current model of industrial forestry. The chapter will also present two propositions that might be called the fundamental theorems of community forestry. These propositions assume that the community can hire talent, mobilize capital, and ensure ethical behaviour roughly as easily as a conventional firm. They say nothing about the adjustment period that might be required to develop the institutional forms and the human capital to handle the more complex problem that maximizing welfare involves.

An implication of the two propositions is that **community forestry is a more general economic form than conventional industrial tenures**: the benefits of conventional tenure are available within a community forestry framework. The converse is not true: many of the benefits of community forestry are not available under conventional forest tenure. For legislators, the implication is that the legislative framework should make community forestry the basic framework within which more limited forms of tenure are available when the entire range of instruments available to a community forest are not needed.

The Economics of Community Forestry versus Forestry Economics

The first essential point to make is that community forestry is not simply a sub-field of forest economics. Forest economics is an established field in economics. A great deal of standard forest economics will apply to community forestry. Community forestry, however, introduces fundamentally new issues that cannot be dealt with using the tools that characterize forest economics. The special features arise out of the combination of the terms "community" and "forestry." As a result, community forestry requires specific tools from distributional theory, economic development, human capital theory, information economics, public goods theory, principal agent theory, and the theory of social capital, as well as rent theory and many of the tools of cost-benefit analysis and environmental economics already part of forest economics.[1] Figure 12.1 shows some of the sub-fields in economics that would be counted as topics in the economics of community forestry. Each of these sub-fields in economics considers the effect of variables that the community might legitimately control or influence.

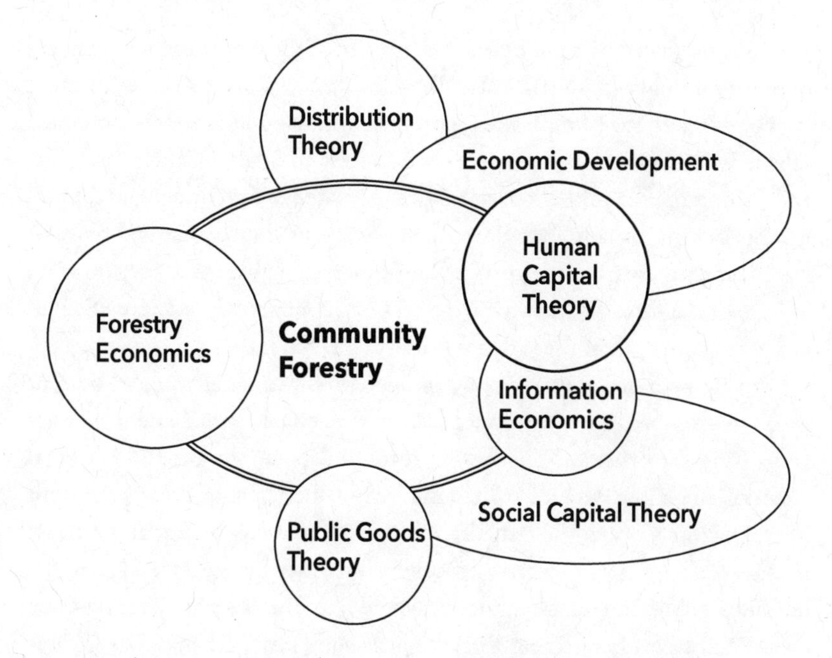

Figure 12.1. Community Forestry and Several Strongly Related Fields.

Issues and Instruments in Community Forestry

It is useful, perhaps, to describe some of the economic theory that is relevant for the economics of community forestry and to identify some of the instruments, such as worker management, available to community forests that might be less accessible or of less interest to firms operating under the current tenure system.

Addressing Public Goods, Externalities, and Joint Production

Publicness and Community Forestry

The theory of public goods would certainly take up a chapter in any book on the economics of community forestry. A public good is an asset or good that anyone can enjoy without reducing the amount available to others. The CBC is a public good. A local public good, like a street lamp, has a more limited availability. Public goods have unusual economic properties. When one person uses a public good it is still available for others. This is quite different from "private goods" like apples or two-by-fours, which are consumed in use. In addition, it is generally hard to exclude people from a public good. The central observation in public goods theory is the famous Samuelson condition, that the value produced by a public good is the sum of the benefits to all members of the community (Samuelson 1954).

Forest roads, for example, are inputs to the production of a private product, salable wood, but they also may provide access to lakes, hunting, camping, and fishing opportunities, as well as scenery, shelter for cottages, and tourism operations, usually without reducing value of the harvest and without significantly increasing the costs of forest management. These uses draw on aspects of the forest with varying degrees of publicness. To achieve efficient use of the forest, all of these benefits must be taken into account.

Private operators cannot easily "monetize" these benefits under current forest tenures because they cannot collect a fee equal to the benefit they are generating for others at the margin (economists emphasize the marginal effect of an action, which is to say the size of the effect of the current unit of production). Since the benefits to others cannot be monetized, private managers don't count them. As a result, private managers provide less of the unvalued public benefits than society would like. This is one way markets are often inefficient. Privatizing forest land is sometimes proposed as a solution, but it would be necessary to privatize the public benefits as well to get managers to fully consider them, and that seems to be a step in the wrong direction. Governments can also regulate, to try to force private managers to supply the correct amount of public goods—for example to open roads for recreational use.

Externalities and Community Forestry

The related concept of an externality has wide application in forestry management. The theory of externalities is generally credited to A.C. Pigou (1920), who applied the concept to, among other areas, road congestion. It is central in Gordon's (1954) analysis of open-access fisheries, which can also be applied to community forestry management.

An externality, or "external effect," is present whenever a decision maker's choices affect others and the external effect is not taken into account in making the decision. As with public goods, the sum of the benefits and costs to all members of the community must be counted. Economic efficiency is achieved when the sum of the marginal benefits is set equal to the marginal costs. Since private producers are not paid for positive externalities or charged for negative externalities, they will ignore these consequences and will in general oversupply or undersupply the product or activity in question. If harvest affects water quality downstream, for example, the harvester may ignore that cost, since it is borne by others, and cut too close to watercourses. It is easily seen that a community forestry organization is likely to consider a wider range of effects than is a private forestry company because more of the people affected are involved in decision making. Many externalities for the firm are "internalized" in the move to community forestry.

Barnett and Yandel (2009: 131) observe, as have many others, that "all externality problems can be considered as property rights problems." Tenure, as a property right, is seriously misaligned with the full range of costs and benefits of forestry. Reassigning rights so that someone has a stake in maximizing the net social benefit is a common prescription in the face of environmental problems, as described in Scott Gordon's famous paper (Gordon 1954: 125) on open-access fisheries popularized much later by Hardin in "The Tragedy of the Commons" (Hardin 1968: 1244). The proposition underlying community forestry is that the assignment of property rights to the community will increase the economic efficiency of forest use.

The property rights problem has another important implication. Zhang (2001: 198) argues that "the costs associated with defining, protecting and transferring property rights in trees, land and products—the 'transaction costs'—are much higher in forestry than in agriculture." It has been argued that firms exist in order to minimize the cost of market transactions that involve invoicing and collecting payment, writing contracts, ensuring that contracts are fulfilled, and other activities that can be handled expeditiously inside the command and control structure of a conventional firm (Commons 1931). Because ownership is complex and many interests are involved in forestry, Vega and Keenan (2014: 2). argue, as does Zhang (2001), that community forestry can also economize on transaction costs.

Joint Production and Community Forestry

Even the simplest possible forest, one with even-aged stands of a single species, produces many outputs. Mixed-species, mixed-aged forests make the problem even more complex. While wood is easily the most important revenue generator in this conventional economy, the forest also provides a range of ecosystem goods and services, such as water storage and filtration, carbon sequestration services, canoe routes, and owl habitat (see also Palmer and Smith, Chapter 2). Harvesters and mill owners do not earn revenue for the services of the forests in cleaning water or sequestering carbon, and as a result they will take these products into consideration only out of good will or because of regulation. Economists speak of joint production in cases like this. Baumgärtner et al.

(2006) argue that "joint production" of this sort is the structural cause behind modern-day environmental problems.

The joint production problem is related to externalities and publicness, but it is perhaps the most fundamental of the three. The central observation in the theory of joint production is that the sum of the values of all the products of a firm minus the sum of all costs should be maximized, and not the net values of any subset of production. As with public goods and externalities, efficiency requires a summation of marginal costs and benefits.

A model with just two goods provides a good deal of insight into the problem of joint production. In Figure 12.2, efficient combinations of timber and recreation that can be produced by a local forest are shown as the dark curved line (known as a "production possibility frontier" or PPF). Combinations below the curve are possible but not efficient. Combinations above and to the right of the curve are not possible. The problem in assigning property rights is to choose a pattern of rights in which the people involved naturally and reliably choose the point on the PPF that is best for society as a whole.

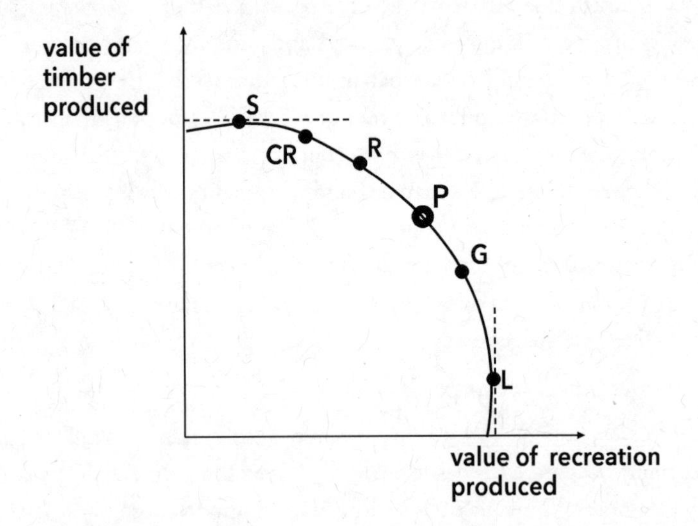

Figure 12.2. Political Economy of a Multi-product, Multi-interest Forest.

The dotted horizontal and vertical lines that are tangent to the Forest Production Possibilities Frontier in Figure 12.2 show the largest values possible for a lodge owner and a shareholder in a timber firm. A lodge owner who is only interested in recreational services would choose *L*. The shareholder of a timber firm is only interested in timber harvest, and would choose *S*. Taking all benefits and costs, including externalities and potential joint product, into account, society might prefer another point, say *P*.

As previously discussed by Palmer and Smith in Chapter 2, the current tenure system is rooted in a particular historical context. Sometime in the seventeenth century the Crown adopted a policy of allocating forest rights to harvesters in exchange for Crown revenue. The forests at that point were essentially uninhabited from the Crown's point of view. Only the timber values and the Crown revenue mattered. As Stephen Harvey observes in Chapter 3, the interests of the Crown and the timber companies were well aligned. Both desired to be at *S*, as far to the top of the graph as possible.

Now that a market for recreational services has emerged, the Crown might prefer to shift production to the point on the curve that maximizes the sum of the revenues from both markets. If, for example, recreational services from a hectare were selling for twice the price that timber was selling for, the Crown (or government) would choose *G*.

Faced with an incumbent forestry company, however, the Crown might attempt to regulate the firm, requiring the firm produce at *R*. The firm would of course resist moving to *R*, and regulation can often be evaded. At a small cost, however, represented by a small downward shift from *S*, the company might shift the output mix horizontally to a compromise at *CR*. This provides a considerable gain to the recreation operator.

A small concession like this demonstrates good corporate citizenship and often yields significant political capital. In the regulatory environment, political capital often takes the form of recognition or "credit" for undertaking a costly action that benefits another. This credit can be used to justify not taking on another cost that the regulators might demand. For example, a company might argue something like this: "Since we have already moved the road that we planned in order to accommodate the Sunny Times Lodge, you can't ask us to leave valuable timber along that ridge on the off-chance that it might support some salamanders." If the owner of the forest company owned the lodge, she would also choose a point close to *G*. The Crown's preferred solution could be achieved by privatizing the forest entirely, which is the solution that many economists might suggest.

The main argument for privatizing forests, in fact, is that a private owner with the right to exploit every aspect of the forest might take all of the values into account. Although in principle the lodge owner–shareholder might want exactly the mix that is best for society, it is much more likely that a private owner would ignore so-called "downstream" effects, the value of training, and other benefits, requiring additional regulation to ensure the resource is used efficiently.

The important point, however, is that the economic values for different players are filtered through the rights they have. Theory holds that neither the lodge owner nor the forestry company should be entrusted with the entire forest. Government may attempt to reconcile the various interests. Having assigned legal rights, government may also impose regulations. The outcome is determined by political realities as well as economic considerations.

Community forestry has the potential both to represent the public interest in a locally sensitive manner and to resolve some of the conflicts inherent in the joint product, high externality nature of forestry. A community forest organization is likely to choose a point near *P*. In doing so it takes on some of the roles of higher level government. It may or may not involve direct production of timber or recreational services in this case, since if those activities can be assigned by government to a private producer, they can also be assigned to a private producer by a community forest organization. The significant difference is the choice of an output combination closer to the socially optimal output.

Human Capital and the Community Forest

Human capital is not an incidental product of human communities—it is the primary product. In a general sense all economic activity is about producing and reproducing people. The fact that communities fund schools indicates that creating human capital is publicly understood as an essential social activity, and possibly one too important to entrust to the private sector. For private producers, however, human capital is an incidental joint product for which they are not paid. The reason for the perceived shortage of skilled trades in Canada is simply that employers are reluctant to invest in skills development because they cannot reliably capture the return on skills developed. Forestry companies with conventional tenure might have some interest in producing human capital, but their interest is obviously less direct than is the interest of the community as a whole. Responsibility for human capital formation has therefore increasingly fallen to the public sector.

Communities, on the other hand, have historically been responsible for local education, and families in a community have a direct stake in effective education. A community forestry organization would be more likely than a large firm to look for ways to combine forestry operations and local education, hoping to improve both. It might, for example, encourage the study of silviculture in the local schools, employ local students in operations, support schools in doing silvicultural research, and accept the responsibility for supervising apprenticeships. Furthermore, since community forest organizations are easily seen as representatives of the public interest with respect to these issues, and forest companies can rarely make this claim, community forestry organizations should be freer than purely commercial operators to experiment with this kind of integration.

Community Forestry and Social Capital

Community building and network development are not normally considered forest products, but in fact any community engaged in production creates considerable social capital. Social capital, like human capital, is itself a product of economic and social production. In the context of this chapter, social capital is the term for a collection of capacities that are expressed through social networks.

There are two common conceptions of social capital. A sociological emphasis on stratification, class, privilege, and power naturally tends to theorize social capital as a private asset from which some are excluded. The concept of social capital as an individual acquisition is associated with Bourdieu (1986). In economic theory, social capital is more often seen as a public or local public good. Economists tend to focus on the productivity of the social asset and the costs of producing it: "the network of relationships is the product of investment strategies, individual or collective, consciously or unconsciously aimed at establishing or reproducing social relationships that are directly usable in the short or long term" (Bourdieu 1986: 52).

Community organizations often create shared capital, including social capital. They invest in public works, develop schools, and support sports organizations. Community forest organizations will similarly value social capital in both forms. They will have opportunities to provide members of the community with connections and practice in working with others that expand individual's social capacity, and they will have strong incentives to improve the communal institutions and networks. These investments would provide little return to the owners of a conventional forestry firm. Economic theory therefore predicts that there will be under-investment in social capital under the current tenure regime.

Community Forestry and Economic Development

The role of community forestry in economic and social development has received considerable attention. Sunderlin (2006), for example, considers the role of community forestry in the alleviation of poverty. Bray et al. (2003, 2006) apply concepts from common property theory and social and natural capital to analyze the emergence and operation of "a very large sector of community forests managed for the commercial production of timber."

A concern for human development provides an answer to a question raised by Luckert (1999). Why bother maintaining forestry communities that developed to serve an industrial mode that no longer needs them? The answer is not that the communities in some sense deserve special exemption from the forces of the market. A sentimental concern for a dying way of life may be a poor guide to forest policy. A more convincing argument rests on the value of these communities to the rest of society. Forest-based communities can be a place where human talents are developed while providing indispensable environmental and economic services.

Furthermore, climate change, carbon sequestration programs, and rising wood values are likely to require enhanced levels of forest stewardship. The people in these communities would then have an expanded role in caring for the forests of their regions. The interests of the forest communities are already intrinsically aligned, not with current social policy but with a more ambitious and forward-looking forestry policy (such as those courses of policy outlined by Palmer and Smith, Chapter 2). They just need the social policy to catch up.

Finally, if Canada's Crown forest tenure systems are reformed to encourage the development of value-added forestry—both pre- and post-harvest—then forest-based communities will be the basis of wealth creation to the benefit of the larger society. Experts believe that adding value to wood is the only possible direction for increasing employment and wealth in forest-based regions (Moazzami 2006).

The Fundamental Theorems of Community Forestry

This section presents two propositions that might be called the fundamental theorems of community forestry. (The propositions are presented more formally in Robinson [2017].) Economists will recognize that the "fundamental theorems" are restatements of basic economic principles. The "theorems" simply present formally what many already believe about the potential advantages of community forestry over the current tenures system.

Expanding the Set of Choice Variables

As a legitimate representative of the public interest, a community forest organization would be permitted to influence variables not available to private operators. A basic result in economic theory states that relaxing a constraint cannot hurt. The result implies you cannot do worse if you have more tools, and you might do better. In other words, if you control both x and y, you cannot do worse than if you only control x.

The logic is simple enough: it is always possible to not use an extra tool, so it is always possible to do as well as you would if you did not have the tool. In principle, then, the wider the range of powers available to a managing entity, the better the outcomes. The conclusion follows immediately, but it helps to be slightly more formal.

We can express the economic problem for a forestry company the following way:

> ***maximize profits*** *given the features of the forest, the labour market, the technology available, and subject to any enforceable rules.*

Some activities, such as lobbying, have a price but do not earn revenue directly. Some variables are controlled by the community but not by the firm. These might include speed limits, railroad schedules, power costs, water supply, or when high-school students are available to work. There are other variables that the community cares about that are not part of the list of variables the firm cares about. These might include water quality, educational attainment, and recreational access to the forests.

The technology available to the firm imposes certain constraints. Taken as a whole, the technology specifies all the combinations of labour, trucks, road building, and fuel, for example, capable of producing any given combination of salable timber in any section of the forest licence. It might specify how much labour is needed for various types of trucks, and the quality of road construction needed to operate various truck sizes. Similarly, firms are subject to a set of rules or regulations. One might specify that for every tree of type A cut, ten seedlings must be planted.

Let's assume that a community forest company is only concerned about profits. It would have the same rights to the same forest as does the conventional forestry company, but a community forest organization would typically also have a social licence to make decisions about some additional social and environmental variables. We can express the economic problem for a forestry company the following way:

> ***maximize profits*** *given the features of the forest, the labour market, the technology available, and subject to any enforceable rules,* ***taking advantage of any additional variables*** *that the community controls.*

Since a community can influence extra variables that are not available to the forestry company, the community forestry organization will normally be able to achieve higher profits. How much higher depends on a variety of conditions, but the basic proposition is a mathematical truism. Starting from the private profit–maximizing point, the managers of the community forest can manipulate some variables not available to the company to earn even more profit. A community forestry company might, for example, sell garbage services to the local municipality, allowing the community forest company to achieve some economies of scale while cutting costs for the community. The community could in principle contract with a company that it didn't own, but because the books of a private company are generally not open, transaction and monitoring costs would probably be higher, liability questions would be harder to solve, and company effort would be spent in rent seeking. Although communities and companies often assist each other, the inability of a private forestry company to control the community variables suggests that many opportunities will be missed. A lot of "small change" will be left on the table.

Combining the above observations, we have the **first fundamental theorem of community forestry**:

> *A community forest operation will be able to generate at least as much profit as a private forestry company can achieve if the community forest has control of additional variables as a result of its status as a community organization.*

Expanding the Set of Variables that Count

Including public goods, externalities, joint production, and human and social capital leads us to the second fundamental proposition about community forestry: the community will consider a larger set of variables than a conventional forestry company. Assume the community cares about variables that do not contribute to the profits of the forestry operation. The community forestry organization would then maximize the combination of profits from its forestry operation plus the benefits from the additional variables.

It is convenient to introduce another mathematical truism at this point: a company trying to maximize profits will not generally achieve as much community welfare as a community that is trying to maximize community welfare. In general it will do worse. Therefore, the **second fundamental theorem of community forestry** is:

> *A community forest operation will achieve the goals of the community more reliably than a private forestry company if the community is interested in more than profits for stockholders.*

There is an implicit assumption here that the community can hire talent, mobilize capital, and ensure ethical behaviour roughly as easily as a conventional firm. Furthermore, the proposition says nothing about the adjustment period that might be required to develop the institutional forms and the human capital to handle the more complex problem that maximizing welfare involves.

Conclusions

Current forestry practice in Canada is based on institutional forms that are centuries old—that pre-exist the country itself. They have co-evolved with technologies and labour markets and must be understood as mature systems. Community forestry, although it has deep roots, is still a series of small, spontaneous experiments in Canada. Comparing examples in British Columbia and Mexico, Davis (2008: 11) summarized the state of community forestry: "While community forestry in British Columbia has achieved laudable economic goals, it is still a diverse and emerging type of tenure." Community forestry in Ontario is still less developed. Earlier chapters by Palmer and Smith, Chapter 2 and MacLellan and Duinker, Chapter 6 illustrate that community forestry in different Canadian provinces truly remains in its infancy.

If community forestry were to become widespread, a wide range of associated institutions and practices would inevitably develop. Financial institutions would have to learn to deal with community forestry organizations. Schools of forestry would train foresters to work with small community forest licences, as is happening in British Columbia, where a provincial community forestry program has developed a market for managers with experience in leadership in democratic organizations (as described by Bullock, Broad, Palmer, and Smith in the Introduction to this volume). Furthermore, technologies would evolve to meet the demand for more sophisticated management.

Attempts to compare the economic efficiency of community forestry and conventional tenures based on existing examples are like pitting a baby against a lumberjack in a tree felling contest. The conclusions are simply not informative given the vast disparity of experiences and understanding between conventional industrial forestry and community forestry.

It has been shown in this chapter, however, that even before community forestry is widespread in Canada and before institutional forms are well developed, it is possible to draw conclusions about the economic potential of community forestry. The economics of community forestry, drawing on "first principles," the most basic and least disputed propositions in economics, makes it clear that there will be many situations in which community forestry clearly dominates more conventional forms of tenure in strict economic terms. Future opposition to community forestry will need to be backed with evidence, since theory provides little or no reason to expect community forestry to be less efficient than conventional forestry and strong reasons to expect community forestry to be more efficient.

An implication of the two theorems presented here is that **community forestry is a more general form than conventional industrial tenures**: the benefits of conventional tenure are available within a community forestry framework. The converse is not true: many of the benefits of community forestry are not available under conventional forest tenure. For legislators, the implication is that the legislative framework should make community forestry the basic framework within which more limited forms of tenure are available when the entire range of instruments available to a community forest are not needed.

It is important to understand this point. Community forestry is not about restricting the wood supply to conventional mills in any way. Community forestry represents an expansion of possibilities rather than a set of restrictions. An industrial forestry company with conventional tenure has a strictly limited right to harvest some part of a specified region, usually accompanied by additional requirements, such as reforestation. The limitations are required because there are significant principal–agent problems. Principal–agent problems arise when the incentives for an agent, in this case the company, differ from those of the principal, in this case the public. In many cases, according to Adam Smith (1904: 456), a company may be "led by an invisible hand to promote an end which was no part of his intention" (i.e., to serve the needs of the public). In more complex cases, where there are benefits and costs for people who do not own the company, the market does not automatically align the interests of the company with those of the public (see Palmer and Smith, Chapter 2). Forestry is such a case.

In such cases regulation may solve the problem, but only if it is possible to devise a comprehensible, flexible, and cheaply enforceable set of rules. Regulation, however, introduces additional problems and additional costs, including the cost of evasion, the costs of oversight, bureaucratic delays, and costly decision making, inflexibility, and regulatory capture. All of these undermine public trust in the conventional system and increase costs for private firms. Community forestry resolves some of these problems by making a representative of the public—the local community—the operator, thus aligning the interests of the public and the operator. In this way, community forestry reduces costs by decentralizing decisions.

In *Why Nations Fail: The Origins of Power, Prosperity, and Poverty*, Acemoglu and Robinson (2012) argue that it is economic and political institutions that are the foundation of economic success: they create incentives for wealth creation, reward innovation, and allow widespread participation in economic opportunities. Community forestry is an economic and political institution that provides new incentives for innovation and mobilizes human capital in a way that industrial forestry cannot.

Notes

1 See, for example, Bray et al. (2006), for an example of applying common property and social capital theory to the emerging Mexican community forestry sector.

References

Acemoglu, D., and J.A. Robinson. 2012. *Why Nations Fail: The Origins of Power, Prosperity, and Poverty*. New York: Crown Business.

Arnolds, J.E.M. 2001. *Forests and People: 25 Years of Community Forestry*. Food and Agriculture Organization of the United Nations. http://www.fao.org/docrep/012/y2661e/y2661e00.htm.

Barnett, A.H., and B. Yandle. 2009. "The End of the Externality Revolution." *Social Philosophy and Policy* 26 (2): 130–50.

Baumgärtner, S., M. Faber, and J. Schiller. 2006. *Joint Production and Responsibility in Ecological Economics on the Foundations of Environmental Policy*. Northampton: Edward Elgar.

Bourdieu, P., 1986. "The Forms of Capital." In *Handbook of Theory and Research for the Sociology of Education*, edited by J. Richardson, 241–58. New York: Greenwood.

Bray, D.B., C. Antinori, and J.M. Torres-Rojo. 2006. "The Mexican Model of Community Forest Management: The Role of Agrarian Policy, Forest Policy and Entrepreneurial Organization." *Forest Policy and Economics* 8: 470–84.

Bray, D.B., L. Merino-Pérez, P. Negreros-Castillo, G. Segura-Warnholtz, J.M. Torres-Rojo, and H.F.M. Vester. 2003. "Mexico's Community-Managed Forests as a Global Model for Sustainable Landscapes." *Conservation Biology* 17 (3): 672–77.

Commons, J.R. 1931. "Institutional Economics." *American Economic Review* 21: 648–57.

Davis, E.J. 2008. "New Promises, New Possibilities? Comparing community forestry in Canada and Mexico." *British Columbia Journal of Ecosystems and Management* 9 (2): 11–25.

Devkota, R. 2010. *Interests and Powers as Drivers of Community Forestry: A Case Study of Nepal.* Goettingen: University Press Goettingen.

Gordon, H.S. 1954. "The Economic Theory of a Common-Property Resource: The Fishery." *Journal of Political Economy* 62 (2): 124–42.

Hardin, G. 1968. "The Tragedy of the Commons." *Science* 162 (3859): 1243–48.

Luckert, M.K. 1999. "Are Community Forests the Key to Sustainable Forest Management? Some Economic Considerations." *Forestry Chronicle* 75 (5): 789–92.

Moazzami, B. 2006. *An Economic Impact Analysis of the Northwestern Ontario Forest Sector*. North-Western Ontario Forest Council, Department of Economics.

Pigou, A.C. 1920. *The Economics of Welfare*. London: Macmillan.

Robinson, David. 2017. *The Economic Theory of Community Forestry*. Oxford: Routledge.

Samuelson, P.A. 1954. "The Theory of Public Expenditure." *Review of Economics and Statistics* 36: 386–89.

Schusser, C. 2012. "Community Forestry: A Namibian Case Study." In *Moving Forward With Forest Governance*, edited by G. Broekhoven, H. Svanije, and S. von Scheliha, 213–21. Wageningen: Trobenbos International.

Smith, Adam. 1904. *An Inquiry into the Nature and Causes of the Wealth of Nations*. Edited by Edwin Cannan. London: Methuen. Library of Economics and Liberty. Accessed 23 June 2015. http://www.econlib.org/library/Smith/smWN13.html.

Statistics Canada. 2015. *Table 228-0059 – Merchandise imports and exports, customs and balance of payments basis for all countries, by seasonal adjustment and North American Product Classification System (NAPCS), monthly (dollars)*, CANSIM (database). Ottawa. Accessed 19 June 2015. http://www5.statcan.gc.ca/cansim/a26?lang=eng&id=2280059.

Sunderlin, W.D. 2006. "Poverty Alleviation through Community Forestry in Cambodia, Laos, and Vietnam: An Assessment of the Potential." *Forest Policy and Economics* 8 (4): 386–96.

Vega, D.C., and R.J. Keenan. 2014. "Transaction Cost Theory of the Firm and Community Forestry Enterprises." *Forest Policy and Economics* 42: 1–7.

Zhang, Y. 2001. "Economics of Transaction Costs Saving Forestry." *Ecological Economics* 36: 197–204.

CONCLUSION

Lessons from Community Forestry Practice, Research, and Advocacy

Ryan Bullock, Gayle Broad, Lynn Palmer, and M.A. (Peggy) Smith

The initiatives explored in this volume demonstrate that capacity exists among forest communities to drive partnership development, innovation, economic diversification, institutional advancement, and environmental stewardship. Communities, whether they are on the "leading edge" of new forest development and governance models or working within somewhat conventional arrangements, can learn from each other, as can policy makers and researchers. However, these learning communities of practice need resources and a space through which to disseminate knowledge and resources in support of their collaboration. As presented in the previous chapters, Canada holds decades of experience with local involvement in forestry, provincial community forest programs, and multi-party collaboration. Together the chapters presented here tie together concepts of relevance to community forestry with actions and perspectives "on the ground." In this conclusion we distill lessons from communities across Canada's many forest regions.

Lesson 1: There Is a Need and Opportunity to Do Forest Business and Development Differently Using Collaborative Governance

There are many approaches to community and forest-based development unfolding that fit with the community forest principles of local decision making, values, and benefits (Wyatt et al. 2013). In part, the diversity of models used to pursue forest development and governance goals is also required to address the cultural and ecological diversity across Canada's vast forest regions (see Rowe 1972). As presented by Harvey in Chapter 3, governments often assess a wide range of models for involvement prior to experimenting with some version of community forestry. While community forestry is one approach, land-based development on Crown land, First Nation reserve land management, and the like, represent common resource management options that can perhaps serve to promote the values and standards expected by the full "range of publics" (Harrington et al. 2008: 201) or the citizens and their communities who are the actual owners of Canada's forest lands.

There are longer-term forces of change at work driving forest sector transition. For example, Howlett (2001) suggests that state-level acknowledgement of Aboriginal rights has actually encouraged innovation and experimentation with forms of multi-level forest governance. Public attitudes and perspectives continue to evolve with respect to Indigenous peoples toward a more supportive and conciliatory position (Coates and Crowley 2013). However, the resounding demands of Indigenous peoples themselves are paramount to them becoming more involved in forest governance (McGregor 2011). As illustrated by Harvey (Chapter 3 in this volume), public values regarding forests also continue to change. Increasingly, Indigenous and non-Indigenous communities see that they share more in common in terms of interests and values than once believed. The emergence of new cross-cultural and reconciliatory arrangements in forestry, but also other resource sectors, is a testament to this fact. Casimirri and Kant (Chapter 4), Lachance (Chapter 5), and Murphy, Chretien, and Morin (Chapter 11) all illustrate how collaborative initiatives at work in Canadian forest landscapes must engage a diversity of values that requires new and flexible management approaches. Bullock, Teitelbaum, and Lawler (Chapter 1) also illustrate how diverse networked approaches to forest governance are being advanced by different combinations of federal, provincial, municipal, and Indigenous governments, together with private companies and NGOs.

Much of the progress and learning that has occurred regarding community forestry and related initiatives is the result of collaborations involving willing but also sometimes reluctant groups. As seen in the chapters of this volume, collaborative multi-level networking involving different actors enables new ways of working together. It is especially important for groups of people to get a chance to meet regularly with each other, build relationships, and identify common objectives prior to formalization of collaborative arrangements. Changing conditions in the forest sector and forest communities have motivated various groups to reach out to one another. This has generated new arrangements for conventional and "new" players to engage in forest governance and development (NAFA 2015).

First Nations in particular have been instrumental in encouraging the rebuilding of relationships as well as directing policy and management agendas. Casimirri and Kant (Chapter 4) illustrate that government, industry, and First Nation community relationships have a better chance of being successful when they are supported by participants' common commitment to a given place and community. Their research shows that a number of deliberate actions can be taken to help support cross-cultural and cross-level relations, including developing both formal and informal social networks, as well as establishing intercultural liaisons so participants can get to know one another but also deliberate in a structured, meaningful manner. Building trust, understanding, and respect among groups is more likely to support mutual goals and recognition of responsibilities. As Lachance points out in Chapter 5, ambitious First Nations can use forestry as an entry point for involvement in many kinds of land use development initiatives with a variety of partners. However, these new relationships require new models for working together, which has driven innovation.

Sudden changes and surprises are also possible, of course. Certainly the forest sector crash in the past decade that saw 130,000 Canadian forestry jobs lost (CCFM 2015b) provided a major impetus for exploring new ways of working together. The Canadian Council of Forest Ministers (CCFM) called the need for innovation in Canada's forest sector "critical to the transformation necessary for sector renewal" (CCFM 2015b: 2). At the same time, according to Hansen et al.'s (2014: 1346) research review on North American forest sector innovation, in general "forest industry managers do not see their operations as highly innovative." Moves to make forestry more open to smaller businesses and organizations, such as reforming tenure systems to allow new entrants into the sector, are meant to help diversify and encourage novel thinking and practice. After all, forest communities, economies and cultures can become entrenched in the past (Bullock 2013). Power struggles, pressure, conflict, and negotiations involving conventional players and new advocacy groups are needed to promote transformational change (Geels and Schot 2007). Both social and technological innovation is needed for regional system transformation, and the definitions and sources of each do not lie in a single group or domain (Cooke et al. 1997), suggesting that broad involvement and collaboration are actually necessary for change. Murphy, Chretien, and Morin (Chapter 11) illustrate how blending traditional practices and Indigenous knowledge with recent technological innovations and marketing approaches contributes to non-timber forest sector diversification. This is echoed by Egunyu and Reed (Chapter 9), who point out that broadening participation in forest management can enable learning and help to create new opportunities for non-conventional actors to build new skills and knowledge that are in line with local objectives. For these and other reasons, not only is collaboration in forest governance and development now considered to be beneficial but, according to the Canadian Council of Forest Ministers' 2015 *Kenora Declaration on Forest Innovation*, collaboration is essential to sparking innovation and experimentation in the forest sector (CCFM 2015a). Gunter and Mulkey (Chapter 8) and also MacLellan and Duinker (Chapter 6) demonstrate how the value of regional networking for collaboration, when guided by commonly accepted principles, can facilitate sharing of lessons from practice and policy improvement. Thus, the chapters of this book as well as previous literature show how community forests can help enable collaboration and therefore innovation.

Lesson 2: New System Opportunities Require New Policies and Adaptive Measures

There is evidence of a growing acceptance of the need for institutional reform and to design appropriate policies to enable community forests in practice. The provincial forest tenure systems across Canada are out of date and no longer suited to supporting communities or companies experiencing major economic, social, and environmental shifts (Haley and Nelson 2007). Communities, industry, and governments recognize that there is a need for change. As Robinson (Chapter 12) points out, Canada's existing policies are not conducive to enabling community involvement, and in fact they are underdeveloped and community forestry is only partly theorized at best. Lack of awareness about community forest concepts and practice among specialists, managers, and lay people alike has contributed to slow adoption and development (and indeed even hampered the belief that community forestry *can* work). Despite the relatively long history of ideas and support for community forestry in North America, community forestry traditions and ideas remain at the margins (Baker and Kusel 2003). Establishing the idea in the mindset of citizens, forestry practitioners and schools, business people, and policy makers remains an important step toward wider acceptance, real support, and, ultimately, change. As the chapters in this volume describe, the challenges of creating awareness and encouraging adoption can also be seen as problems of coordination, promotion, and timing among community forestry supporters.

As demonstrated by Gunter and Mulkey (Chapter 8) as well as MacLellan and Duinker (Chapter 6), Palmer and Smith (Chapter 2), and Harvey (Chapter 3), policy development involving policy makers, practitioners, and researchers can effectively customize policy for regional settings to enable new arrangements. A networked approach can have positive results through producing collaborative research and advocacy initiatives that are informed by and shape practice, and in turn advance policy. Recent experiences in British Columbia with the British Columbia Community Forest Association (BCCFA), Nova Scotia with the Nova Forest Alliance and Provincial Community Forestry Advisory Board, and Ontario with the Northern Ontario Sustainable Communities Partnership speak to the power of collaborative multi-party and multi-level networked arrangements to address needs arising from complex, changing settings. Regional networks aimed at policy development and advocacy, when supported by key provincial bodies where jurisdiction over natural resources lies, can be quite effective forums for producing meaningful change. In addition to providing "essential services" by coordinating and connecting forest interests around timely common issues, regional entities such as the BCCFA have arguably had more success influencing policy than the former well-resourced federal research networks (e.g., Canadian Model Forest Network). Admittedly, these regional advocacy networks have not been pressed to meet the full research mandate held by other federal forest research networks (see Klenk et al. 2010). Nonetheless, creation of additional regional forums for ongoing networking would appear to be important for engaging diverse groups in processes of policy innovation and reform; however, blending deliberative activities with supporting research capacity could produce optimal results in terms of developing new institutions suited to current needs. In other words, both multi-party policy dialogues and research projects need to occur in the same network and, most importantly, they need to involve groups from the level where resource jurisdiction lies (i.e., the provincial/territorial levels or Indigenous governments) in order to make direct and meaningful impacts.

In spite of all the advocacy work that has been done and all the change that has occurred in the forest sector, it is still the case that communities themselves remain the primary advocates for increasing community control and benefits from forests. In Chapter 8, Gunter and Mulkey highlight the efforts of long-toiling BC communities to collectively advocate for policy reforms to enable opportunities for forest communities. The grassroots work of communities played a crucial role in advancing community forestry during the 1970s and 1980s (Booth 1998; Bullock et al. 2009). But as pointed out by Palmer and Smith (Chapter 2) as well as Glynn (Chapter 7), community–university connections have also been instrumental in creating forums and space for public dialogue around issues of interest to communities in forest regions. Such forums often bring forward concepts, information, and options that are later presented to governments searching for solutions to social and environmental issues. In particular, this has been true with the University of Victoria (e.g., Burda 1998; Burda 1999; M'Gonigle et al. 2001) and the Northern Ontario Sustainable Communities Partnership linked to Lakehead's Faculty of Natural Resources Management and Laurentian University (see Palmer and Smith, Chapter 2), as well as efforts undertaken at Dalhousie University (see MacLellan 2012) in eastern Canada. In this way, the conceptualization of policies and models for community involvement in forestry has been greatly advanced by community–university partnerships over the past few decades. The space and knowledge exchange opportunities created through these sorts of partnerships represent an important form of extension work that many communities have accessed to advance their own agendas in the forest. Governments wanting to support community forest development could initiate extension programs alongside policy reform processes to enable transition.

Given the affinity of communities and the forests featured in this volume, it is likely that each community-forest system (see Bullock and Reed 2016; Palmer et al. 2016) will continue to co-adapt. Thus, our institutions, value systems, and knowledge must also adapt to suit complex, changing conditions, both locally and globally (Berkes 2010). The structures and processes for environmental resource decision making are not unchangeable, and it is from this viewpoint that Armitage et al. (2012: 246) point out that "governments are not, and in fact cannot be, the most important source of environmental decision making authority. Decision making must now accommodate diverse views, networks and hybrid partnerships among state and non-state actors, and must include opportunities for shared learning." Recognizing the inherent complexity of environmental resource governance calls for adaptive, multi-party policy-making approaches that are too numerous to list here. New policies based on continuous learning are appropriate because, as recent experiences in Nova Scotia (see MacLellan and Duinker, Chapter 6) and Ontario (Palmer and Smith, Chapter 2; Harvey, Chapter 3) show, when policy windows open, those who act on opportunities to provide decision makers with clearly articulated, evidence-based options stand to influence policy decisions the most. The constant evolution of understanding and policy background work are essential qualities of a responsive multi-party forest policy network (Hessing et al. 2005).

Lesson 3: Mobilizing and Resourcing Community Forests Requires New Information, Knowledge, and Skills

The last three decades have seen increasing interest in community forestry in Canada and internationally. This recent attention was predated by a relatively long, but intermittent, history of experiences that provide some "new," or rarely discussed, insights about the current state of affairs. Understanding past events and context is important in terms of creating a shared memory for a movement (Davis and Reed 2013) but also for understanding possible future paths for community involvement in forest governance. We are at a point now with community forests where policy makers, communities, and their partners need not muddle through the establishment of new initiatives. Lessons and designs *are* available.

The concept of community forestry has been redefined for over a century in North America (Dunster 1989). In fact, the individual credited with bringing scientific forestry to North America, Bernhard Fernow, or "that great German forester who came to America in the nineteenth century and created a forestry consciousness in both the United States and Canada" (MacDonald 1935: 140), was an avid supporter of community forestry. As early as 1890, Fernow, then head of the U.S. Department of Agriculture forestry division, published an editorial paper entitled "Communal Forests" in the prominent journal *Garden and Forest*. His strongly worded article advocated for the formation of locally managed forests in every American town to improve use of idle agricultural waste lands and anchor the national economic resource base, with intentions to spread conservation and improve forestry practices: "In fact, no better method of forest-reform could be suggested than by beginning forestry in each town, which as part of the country at large, will influence the movement of the whole. . . . No move in forestry reform could be more promising than the establishment of communal forests" (Fernow 1890: 349). Fernow would later become the founding dean of Canada's first Faculty of Forestry at the University of Toronto (1907) during what was a period of keen public interest in government-grassroots forest conservation. While community forestry would not become a cornerstone of forestry education in Canada, it was important to some early pioneers who helped to build key parts of forestry education and governance infrastructure in Canada. Training skilled individuals to support the growing industry of the day was a central motivation (Apsey et al. 2000). We share the view of Apsey and colleagues (2000: 38) that today's "new working environment[s] calls for a reassessment of roles, as well as a new form and focus of forest education." However, we must also now accept that the social conditions for forest management and development have changed (Haley and Nelson 2007). Thus, forestry training is needed to suit current societal demands, such as those for increasing community involvement, environmental sustainability, and Indigenous inclusion (see Allen and Krogman 2013). As our environment and society evolve, so too must our institutions that structure human-environmental interactions.

Much as was the case 100 years ago, there is currently a need to develop professionals with training suited to community forestry. With tenure reforms in provinces such as BC, Saskatchewan, Ontario, Quebec, and Nova Scotia, the recent increase in community forestry and apparent interest in cross-cultural collaboration suggest there is and will be a need for trained personnel with diverse knowledge and skills of pertinence to community forestry. Robinson theorizes (Chapter 12) that community forestry can create more positions for those with specialized training in community forestry. Further integrating relevant disciplines and curriculum development at universities and colleges could prove timely, not only for the development of community forestry but for the evolution of forestry in general, which, evidently, is working hard to adapt to changing societal values in order to remain socially relevant in Canada. Forestry professionals are concerned about negative public perceptions regarding the discipline and practice of forestry (i.e., negative environmental impacts, male-dominated, low Indigenous involvement) and falling forestry school enrolments (see Hoberg et al. 2003; CIFINRSSC 2006). Establishing community forestry in the academy and professional realms would also help to advance its conceptual basis and provide a common language and methodology (e.g., well-defined and common concepts, principles, and best practices) for community forestry in Canada. To date, the University of British Columbia is the only university offering specialized programming that integrates community and Indigenous forestry (UBC Faculty of Forestry n.d.), though it is creeping into curricula at other schools. As power structures governing forest resources change (i.e., tenure reform), so too must related system components (i.e., curricula, forms of knowledge, decision makers) that provide the capacity to meet the challenges of today's forest economy and socio-political realities.

For Indigenous communities, capacity development is top of mind because it is often a key factor needed to enable involvement and heighten decision-making control and benefits (Wyatt 2008). This refers to not only skills and training for trades but also professional competencies and accreditations needed for the self-determination agenda (Bombay 2010). Long-term, systematic exclusions have created significant capacity gaps with respect to the education and skills training needed by Indigenous peoples to pursue governance and development priorities in, for example, lands and resource business and development (see Coates and Crowley 2013; Parsons and Prest 2003). Such pursuits require fostering competencies in research and development, technological innovation, and public service (RCAP 1996). Lachance (Chapter 5) lays out a model being built by First Nations in northern Ontario to address capacity and knowledge demands in order to support their reasserted role in forest stewardship and development.

Existing research confirms that communities struggle with knowledge and information challenges to integration, namely, when knowledge and information is not readily available or usable because it comes from different sources and in different formats, even languages (or simply has yet to be produced). This is as true for community forestry (Bullock et al. 2009) as it is in other natural resource development settings (see, e.g., Lajoie and Bouchard 2006). A prime example of this challenge comes in a much-overlooked and rarely cited article in the *Forestry Chronicle*, where Auden (1944) fully detailed a plan for a "forest village" in Nipigon, Ontario. His proposal outlined a complete plan for a multi-faceted, community-based enterprise (i.e., timber and value-added; agriculture and wild food harvesting; tourism and recreation; commercial fishing and trapping; energy production; and transportation) based on a public-private partnership and co-operative structures, to be guided by multiple use forest management with the expressed intent of providing community and household self-sufficiency. Auden's proposal was never implemented (for unknown reasons), nor is it ever cited or familiar in professional circles; however, ideas similar to this proposal continue to recirculate, and new initiatives need not "reinvent the wheel."

In recent decades, there are numerous examples of sophisticated plans and feasibility studies for community forests that were never implemented and remain elusive (see Robin B. Clark Inc. 1996 for Malcolm Island, BC; Silva Forest Consultants Ltd. 1998 for Denman Island, BC; Matakala 1994 for Wendaban Stewardship Authority, Temagami, ON). Scholars have also provided detailed "blueprints" for revising provincial policy frameworks to enable community forestry (i.e., Burda 1998). As having information that is readily available and usable is a major challenge, it follows that having all existing information from community-based organizations, consultants, and academics in a searchable, publicly accessible portal could be a valuable resource to community forestry practitioners, researchers, and policy makers. Questions remain about how this could be coordinated, housed, and made interactive.

Clearly there is a need for leadership and common vision to support community forests in Canada. The previous chapters presented numerous examples of such leadership, which in the future could be supported by developing a knowledge network and new institutions for community forests.

Lesson 4: Building Networks Can Help Foster Research and Innovation

While community forestry initiatives at different stages of development value grassroots involvement and co-operation, there is no formal collective structure in Canada for wider coordination of community forestry advocacy and research efforts. It may not be surprising that community forestry has maintained a decidedly local focus given that, in Canada at least, its emergence is owed to grassroots movements and contested landscapes across small town sites and forests in isolated mountain valleys, discrete watersheds, and islands (the BC Gulf Islands and mainland interior). There are also scattered examples in the provincial norths (see Beckley and Korber 1996; Bullock et al. 2009). From a research perspective, the academic, government, and non-government community forestry literature also has inadequately addressed the importance of forming regional and national linkages among community forest organizations, as well as the advantages of integrating ideas, actors, and regional initiatives across the country. The Canadian Model Forest Network (1992–2014) and the Sustainable Forest Management Network (1995–2009) were mandated to do just this; however, both networks were shut down following federal funding cuts, perhaps an indication of the failure of the federal government to provide national direction on forestry issues. Scholars and professionals also debate the direct impacts and relevance of the work done in these networks with respect to whether they were successful at actually changing policies and equipping communities with actionable knowledge (see Ayling 2001; Sinclair and Smith 1999; NRCAN 2006; Klenk et al. 2013). More recent regional networks, such as the British Columbia Community Forest Association (Gunter and Mulkey, Chapter 8) in particular and the Northern Ontario Sustainable Community Partnership (Palmer and Smith, Chapter 2), were formed because of community-level recognition that there is ongoing need for an entity to coordinate and channel information, tools, and partnerships, and to provide technical advice and continuous policy advocacy.

Thinking about community forestry in Canada as a social movement, local debates typically have been about how community forests can be established to secure a land base and access to timber and other opportunities for economic development (see Palmer and Smith, Chapter 2 this volume; Bullock et al. 2009; Wyatt 2008). Less attention has been given to how communities can use forest planning to build regional linkages and participatory capacity. Lachance (Chapter 5), however, indicates that some First Nations in Ontario have found good potential in regional collaboration, which makes sense especially in dispersed rural settings where functional integration and population levels are low. While a focus on "the local" is essential to working out the early phases of emerging community forest initiatives (as in the case of Glynn's overview of local organizing, Chapter 7), overlooking the need to better integrate communities and forests is an important omission in the literature and, certainly, in practice. No studies of community forestry to date have examined the value and potential of regional networking. Current theory and efforts for local control over forest resources seem to miss the point that communities and their partners are, and must be, linked to one another to achieve shared goals and objectives (Bullock and Reed 2016). Sporadic government interest and wavering support has created an unfocused policy agenda and pockets of albeit passionate and dedicated advocates and researchers working in relative isolation. A network is strongly needed.

Creating connections for capacity in resource and forest regions in particular is a topic of ongoing interest among scholars. Bullock and Reed (2016) argue there is a need for closer and more links among the many localized community forestry movements developing across Canada, including stronger linkages among local and national organizations, practitioners, and advocates. Their analysis reflects on factors influencing community roles in forest development to suggest what practical changes could help to generate a well-supported and integrated system of locally managed forests in Canada. These include conscientiously building capacity for local involvement through connecting organizations (i.e., government, non-government, research) sharing resources (data, funding and training opportunities, expertise), and collectively participating in shared spaces for deliberation and collaboration (e.g., public forums, policy processes, and virtual environs). These efforts together provide the apparatus, content, and mobilization potential for ongoing capacity building and collaboration. Maintaining leadership based on experience and memory is also a key concern. An integrated system of communities can greatly benefit from the combined experiences of its members (i.e., developed through both gains and failures) (Davis and Reed 2013). While first-hand experience with community forestry, local organizing, and planning is no doubt ideal, it is not always available. However, the relationships, processes, and collective character (i.e., the confidence and perseverance that come from failing and succeeding together) developed through community-level and regional planning processes are also highly useful in preparing communities for other opportunities that emerge. Many chapters in this book (see, e.g., Palmer and Smith, Chapter 2, Casimirri and Kant, Chapter 4, and Egunyu and Reed, Chapter 9) illustrate the repertoire of local activities that community forestry practitioners, advocates, and researchers engage in, many of which hold transferable lessons for the future. Our hope is that students, researchers, practitioners, and other readers of this book draw upon the many lessons contained in its pages to help grow community forests in Canada.

References

Allan, K., and D. Frank. 1994. "Community Forests in British Columbia – Models that Work." *Forestry Chronicle* 70 (6): 721–24.

Allen, T., and N. Krogman. 2013. "Unheard Voices: Aboriginal Content in Professional Forestry Curriculum." In *Aboriginal Peoples and Forest Lands in Canada,* edited by D. Tindall, R. Trosper and P. Perreault, 279–97. Vancouver: University of British Columbia Press.

Ambus, L., and G. Hoberg. 2011. "The Evolution of Devolution: A Critical Analysis of the Community Forest Agreement in British Columbia." *Society and Natural Resources* 24 (9): 933–50. doi:10.1080/08941920.2010.520078.

Apsey, M., D. Laishlef, V. Nordin, and P. Gilbert. 2000. "The Perpetual Forest: Using Lessons from the Past to Sustain Canada's Forests in the Future." *Forestry Chronicle* 76 (1): 29–53.

Armitage, D., R. de Loë, and R. Plummer. 2012. "Environmental Governance and its Implications for Conservation Practice." *Conservation Letters* 5: 245–55. doi: 10.1111/j.1755 263X.2012.00238.x.

Auden, A., 1944. "Nipigon Forest Village." *Forest Chronicle* 20 (4): 209–61.

Ayling, R. 2001. "Model Forests: A Partnership-Based Approach to Landscape Management." In *Social Learning in Community Forests*, edited by E. Wollenberg, 151–69. Indonesia: CIFOR and the East-West Centre.

Baker, M., and J. Kusel. 2003. *Community Forest in the United States: Learning from the Past, Crafting the Future.* Washington, DC: Island Press.

Beckley, T., and D. Korber. 1996. "Clear Cuts, Conflict, and Co-management: Experiments in Consensus Forest Management in Northwest Saskatchewan." Information report NOR-X -349. Edmonton: Northern Forestry Centre, Natural Resources Canada.

Berkes, F. 2010. "Devolution of Environment and Resources Governance: Trends and Future." *Environmental Conservation* 37 (4): 489–500.

Bombay, H. 2010. *Framework for Aboriginal Capacity-Building in the Forest Sector.* Report for the National Aboriginal Forestry Association. Ottawa.

Booth, A. 1998. "Putting 'Forestry' and 'Community' into First Nations' Resource Management." *Forestry Chronicle* 74 (3): 347–52.

Bullock, R. 2013. "'Mill Town' Identity Crisis: Reframing the Culture of Forest Resource Dependence in Single Industry Towns." In *Social Transformation in Rural Canada: New Insights into Community, Cultures and Collective Action*, edited by J. Parkins and M. Reed, 269–90. Vancouver: University of British Columbia Press.

Bullock, R., K. Hanna, and D.S. Slocombe. 2009. "Learning from Community Forestry Experience: Challenges and Lessons from British Columbia." *Forestry Chronicle* 85 (2): 293–304.

Bullock, R., and M. Reed. 2016. "Towards an Integrated System of Communities and Forests in Canada." In *Community Forestry in Canada: Drawing Lessons from Policy and Practice*, edited by S. Teitelbaum. Vancouver: University of British Columbia Press.

Burda, C. 1998. "Forests in Trust: A Blueprint for Tenure Reform and Community Forestry in British Columbia." *Ecoforestry* May: 12–15.

———. 1999. "Community Forestry: A Discussion Paper." Focus on Our Forests discussion paper.

[CCFM] Canadian Council of Forest Ministers. 2015a. *Kenora Declaration on Forest Innovation*. http://www.ccfm.org/english/coreproducts-innovation.asp?pf=1.

———. 2015b. "Forest Sector Innovation in Canada." White Paper: Opportunities for the Canadian Council of Forest Ministers.

[CIFINRSSC] Canadian Institute of Forestry Interim National Recruitment Strategy Steering Committee. 2006. "The Crisis in Post-Secondary Enrollments in Forestry Programs: A Call to Action for Canada's Future Forestry Professional/ Technical Workforce." A White Paper on Post-Secondary Forestry Recruitment. *Forestry Chronicle* 82 (1): 57–62.

Coates, K., and B. Crowley. 2013. "New Beginnings: How Canada's Natural Resource Wealth Could Re-Shape Relations with Aboriginal People." Macdonald-Laurier Institute.

Cooke, P., M. Gomez Uranga, and G. Etxebarria. 1997. "Regional Innovation Systems: Institutional and Organizational Dimensions." *Research Policy* 26: 475–91.

Davis, E.J., and M. Reed. 2013. "Governing for Transformation and Resilience: The Role of Identity in Renegotiating Roles for Forest-Based Communities of British Columbia's Interior." In *Social Transformation in Rural Canada: New Insights into Community, Cultures and Collective Action*, edited by J. Parkins and M. Reed, 249–68. Vancouver: University of British Columbia Press.

Dunster, J. 1989. "Concepts Underlying a Community Forest." *Forest Planning Canada* 5 (6): 5–13.

Fernow, B. 1890. "Communal Forest." *Garden and Forest* 3: 349.

Geels, F., and J. Schot. 2007. "Typology of Sociotechnical Transition Pathways." *Research Policy* 36: 399–417.

Haley, D., and H. Nelson. 2007. "Has the Time Come to Rethink Canada's Crown Forest Tenure System?" *Forestry Chronicle* 83 (5): 630–41.

Hansen, E., E. Nybakk, and R. Panwar. 2014. "Innovation Insights from North American Forest Sector Research: A Literature Review." *Forests* 5: 1341–55.

Harrington, C., A. Curtis, and R. Black. 2008. "Locating Communities in Natural Resource Management." *Journal of Environmental Policy & Planning*, 10 (2): 199–215.

Hessing, M., M. Howlett, and T. Summerville. 2005. *Canadian Natural Resource and Environmental Management.* 2nd ed. Vancouver: University of British Columbia Press.

Hoberg, G., R. Guy, S. Hinch, R. Kozak, P. McFarlane, and S. Watts. 2003. "Image and Enrollments." *Forum* 10 (6): 22–23.

Howlett, M., 2001. "Policy Venues, Policy Spillovers and Policy Change: The Courts, Aboriginal Rights and British Columbia Forest Policy." In *In Search of Sustainability: British Columbia Forest Policy in the 1990s*, edited by B. Cashore, G. Hoberg, M. Howlett, J. Rayner, and J. Wilson, 120–40. Vancouver: University of British Columbia Press.

Klenk, N., A. Dabros, and G. Hickey. 2010. "Quantifying the Research Impacts of the Sustainable Forest Management Network in the Social Sciences: A Bibliometrics Study." *Canadian Journal of Forest Research* 40: 2248–55. doi:10.1139/X10-138.

Klenk, N., M.G. Reed, G. Lidestav, and J. Carlsson. 2013. "Modes of representation and participation in Model Forests: Dilemmas and implications for networked forms of environmental governance involving Indigenous people." *Environmental Policy and Governance* 23 (3): 161–76. doi:10.1002/eet.1611.

Lajoie, G., and M. Bouchard. 2006. "Native Involvement in Strategic Assessment of Natural Resource Development: The Example of the Crees Living in the Canadian Taiga." *Impact Assessment and Project Appraisal* 24 (3): 211–20.

Macdonald, J.A. 1935. "A Plan for Reforestation Relief Works Projects in Southern Ontario." *Forestry Chronicle* 11 (2): 133–53.

McGregor, D. 2011. "Aboriginal/non-Aboriginal Relations and Sustainable Forest Management in Canada: The Influence of the Royal Commission on Aboriginal Peoples." *Journal of Environmental Management* 92: 300–310.

MacLellan, K. 2012. "Exploring the Pitfalls and Promise of Community Forests in Nova Scotia." Unpublished Master's MREM Report, Dalhousie University.

Matakala, P.W. 1994. *Wendaban Stewardship Authority 20-Year Forest Stewardship Plan, July 1, 1994 to June 31, 2014: A Report Prepared for the Wendaban Stewardship Authority (WSA), Temagami, Ontario.* Temegami: Wendaban Stewardship Authority.

M'Gonigle, M., B. Egan, and L. Ambus. 2001. *The Community Ecosystem Trust: A New Model for Developing Sustainability.* Victoria: POLIS Project on Ecological Governance, University of Victoria.

[NAFA] National Aboriginal Forestry Association. 2015. *Third Report of First Nations-held Forest Tenure in Canada.* Ottawa: NAFA.

[NRCAN] Natural Resources Canada. 2006. *Canada's Model Forest Program (CMFP) – Follow-up and Mid-term Evaluation (E05002), May 2006.* Accessed 3 August 2015.http://www.nrcan.gc.ca/evaluation/reports/2006/886#archived.

Palmer, L., M.A. Smith, and C. Shahi. 2016. "Community Forestry on Crown Land in Northern Ontario: Emerging Paradigm or Localized Anomaly?" In *Community Forestry inCanada: Drawing Lessons from Policy and Practice,* edited by S. Teitelbaum, 94–135. Vancouver: University of British Columbia Press.

Parsons, R., and G. Prest. 2003. "Aboriginal Forestry in Canada." *Forestry Chronicle* 79 (4): 779–84.

Robin B. Clark Inc. 1996. *Malcolm Island Community Forest Tenure Feasibility Study: Final Report.* Vancouver, British Columbia.

Rowe, S. 1972. *Forest Regions of Canada.* Department of the Environment, Canadian Forest Service. No. Fo47-1300. Ottawa.

Silva Forest Consultants Ltd. 1998. *An Ecosystem-Based Assessment of Denman Island.*

Sinclair, A.J., and D. Smith. 1999. "The Model Forest Program in Canada: Building Consensus on Sustainable Forest Management?" *Society and Natural Resources* 12 (2): 121–38.

Teitelbaum, S., and R. Bullock. 2012. "Are Community Forestry Principles at Work in Ontario's County, Municipal, and Conservation Authority Forests?" *Forestry Chronicle* 88 (6): 697–707.

UBC Faculty of Forestry. n.d. "Specialization in Community and Aboriginal Forestry." University of British Columbia. http://www.forestry.ubc.ca/students/undergraduate/prospective/degree-programs/forestry/forest-resources-management/major-specializations/specialization-in-community-and-aboriginal-forestry/.

Wyatt, S. 2008. "First Nations, Forest Lands, and 'Aboriginal Forestry' in Canada: From Exclusion to Comanagement and Beyond." *Canadian Journal of Forest Research* 38: 171–80.

Wyatt, S., J. Fortier, D.C. Natcher, M.A. (Peggy) Smith, and M. Hébert. 2013. "Collaboration between Aboriginal Peoples and the Canadian Forest Sector: A Typology of Arrangements for Establishing Control and Determining Benefits of Forestlands." *Journal of Environmental Management* 115 (30): 21–31. doi.org/10.1016/j.jenvman.2012.10.038.

ACKNOWLEDGEMENTS

As the co-editors of *Growing Community Forests*, it is important to acknowledge that this book is an initiative undertaken by the partners of Community Forests Canada—a national network of organizations and communities committed to improving community and forest sustainability.

We fully acknowledge all the research participants, their communities, organizations, and lands, as well as all of the authors for their willingness to share their stories. We thank the members of the Northern Ontario Sustainable Communities Partnership for contributing to our learning. The University of Winnipeg's Centre for Forest Interdisciplinary Research, Lakehead University's Faculty of Natural Resources Management, and NORDIK Institute at Algoma University provided important venues for discussion and collaboration. We are grateful for the support provided by our many research assistants, including Julia Lawler, Janelle Laing, and Marika Olynyk. They helped to run the events behind this collaborative project, and they very capably helped to coordinate the editing process. We also recognize the two anonymous peer reviewers and editorial staff at the University of Manitoba Press for their constructive feedback that helped to improve the book. Core funding support for this project was provided by the Social Sciences and Humanities Research Council of Canada (SSHRC) and The University of Winnipeg.

Ryan Bullock, Gayle Broad, Lynn Palmer, and M.A. (Peggy) Smith

CONTRIBUTORS

Gayle Broad is an associate professor at Algoma University in the Community Economic and Social Development program and the director of research for NORDIK Institute, a community-based research institute serving northern Ontario. Her research focus has been in the social economy sector, including community resilience, and in participatory action research. A lifelong resident of northern Ontario, Gayle recognizes community forests as an essential part of establishing a sustainable, resilient region.

Ryan Bullock is a Canada Research Chair in Human-Environment Interactions and associate professor in the Department of Environmental Studies and Sciences at the University of Winnipeg. He is also director of the Centre for Forest Interdisciplinary Research. His research focuses on conflict and cross-cultural collaboration in emerging multi-level environmental resource governance systems, as well as community-based research approaches.

Giuliana Casimirri is a consultant, researcher, and program coordinator for several non-profit organizations focused on urban greening, forest conservation, and cross-cultural environmental resource management. Her doctoral research investigated the outcomes and challenges of collaboration efforts between Indigenous and non-Indigenous people for the management of forest resources in northeastern Ontario.

Annette Chretien is a Metis woman from Sudbury, Ontario. Dr. Chretien works as contract academic faculty at Wilfrid Laurier University. Her research is focused on the construction of contemporary Metis identities and the inclusion of Indigenous Knowledge in interdisciplinary research. To that effect, she has published in a variety of research areas, including the duty to consult and inclusion of Metis Indigenous Knowledge (MIK), Indigenous community well-being, and Indigenous education.

Peter Duinker has degrees in agriculture, forest ecology, and forest management. While a professor at Lakehead University in the early 1990s, he was active in research and advocating for new community forests in Ontario. He returned to the community forest theme in recent years, assisting the Government of Nova Scotia with community forest pilot projects.

Felicitas Egunyu is a postdoctoral research fellow in the College of Agriculture at the University of Saskatchewan. She investigates environmental governance, social learning, resource-based communities, and sustainability. Her PhD thesis research examined social learning and collaborative forest management in forest-based communities in Canada and Uganda. Felicitas has conducted research in Canada, Chile, Ethiopia, Kenya, and Uganda.

Tracy Glynn is the Conservation Council of New Brunswick's forest campaign director. She grew up along the river near Upper Miramichi on a mixed farm and with a horse-logging father whose generation, like hers, struggles to make a living in the woods.

Jennifer Gunter is the executive director of the British Columbia Community Forest Association (BCCFA). Since 2002, Jennifer has worked with the board of directors, staff, and membership of the BCCFA to help it grow from an organization of just ten member communities to one with over fifty. Jennifer holds a BA in geography and environmental studies from McGill University and a master's degree in resource management from Simon Fraser University. Her professional interests include community forestry, community economic development, and forest stewardship education.

Stephen Harvey is a senior policy advisor with Ontario's Crown Forests and Lands Branch. Through his thirty-plus years of public service, Stephen has played an instrumental role in policy development to enhance Indigenous and community participation in forestry, as well as advance positive working relationships with resource industries and various forest users. He was involved in the Ontario Royal Commission on the Northern Environment, Ontario Committee on Resource Dependent Communities, and Thunder Bay Waferboard Committee. He led Ontario's community forestry pilot project, receiving an Institute of Public Administration of Canada Gold Award for this endeavour, and he helped develop another community forest in Ontario, Westwind Forest Stewardship, for which he received an Ontario Public Service Amethyst Award.

Shashi Kant is a professor of forest resource economics and management at the Faculty of Forestry and the director, Master of Science in Sustainability Management, at the University of Toronto Mississauga. He specializes in forest resource economics and forest management systems, with an emphasis on economics of sustainable forest management. He is editor-in-chief of a book series, Sustainability, Economics and Natural Resources. He has published six books, including the *Handbook of Forest Resource Economics*, more than seventy-five refereed journal articles, and about forty other refereed articles. He is associate editor of the *Canadian Journal of Forest Research* and the *Journal of Forest Economics*. He has received the Queen's Award for Forestry (2008), the Scientific Achievement Award of Canadian Institute of Forestry (2007), the Scientific Achievement Award of the IUFRO (2005), and the Ontario's Premier Research Excellence Award (2004) for his research.

Colin Lachance is the founding corporate secretary of the Northeast Superior Regional Chiefs' Forum (NSRCF), a political advocacy body created in 2007 to oversee the development of a regional forestry model. He recently became the founding president of Wahkohtowin GP Limited, a for-profit corporation designed to implement the comprehensive community forestry blueprint developed by the NSRCF. Colin received an undergraduate degree in physical geography at Carleton University in 1983 and a master's degree in environmental studies from York University in 1986. He held various positions with the Government of Canada over a sixteen-year period, including national director of environment and natural resources with Indian and Northern Affairs Canada.

Julia Lawler completed a master's degree in science at the University of Winnipeg. Her research interests include community-based resource management, environmental conservation, and forest policy. Julia's thesis research investigated the challenges and outcomes of Indigenous communities' experiences with timber allocations in Manitoba.

Erik Leslie, RPF, is a forestry consultant and the forest manager for the Harrop-Procter Community Co-operative. He is also the president of the British Columbia Community Forest Association. Erik has worked for community organizations, regional and provincial governments, industry, and First Nations on projects from Haida Gwaii to Labrador. He has extensive experience in forestry planning and operations, community consultation, small-business management, and forest certification.

Kris MacLellan holds a master's degree in resource and environmental management from Dalhousie University's School for Resource and Environmental Studies. He assisted the Province of Nova Scotia in the creation of community forests on Crown land and was a contributor to the Halifax Urban Forest Master Plan. Kris works for Minas Energy, where he participates in the development of the province's transition to renewables, including wind energy and the tidal power project in the Bay of Fundy. Kris also serves as a volunteer director at Halifax's Ecology Action Centre.

Grant Morin is a PhD candidate in the Department of Geography at Western University. His research focus is on the environmental and economic impact of the entertainment software sector of Canada's digital economy.

Susan Mulkey is manager of communication and extension with the British Columbia Community Forestry Association. Susan provides outreach and practitioner support for the provincial network of rural, community-based organizations in BC that manage community forests. She has produced a number of publications on community forest governance and organizational development.

Brenda L. Murphy is a tenured professor in the Social and Environmental Justice program at Wilfrid Laurier University, Brantford Campus. Using community-based research and transdisciplinary approaches, her research interests include rural and Indigenous resilience, disasters, and community well-being. Through a focus on maple syrup production, her work also deals with assessing the importance of agro-forestry systems and developing climate change adaptation strategies.

Lynn Palmer is a PhD candidate in forest sciences in the Faculty of Natural Resources Management, Lakehead University, and holds bachelor's and master's forestry degrees. Her PhD research focuses on forest policy to enable community forestry in northern Ontario. She is a founding member of the Northern Ontario Sustainable Communities Partnership, a non-government organization that advocates for community forests in northern Ontario, and Community Forests Canada, a new national network that aims to bridge practice, research, and advocacy for community forests in Canada.

Maureen G. Reed is professor and assistant director of the School of Environment and Sustainability at the University of Saskatchewan. She studies the social dimensions of environmental change, focusing particular attention on how different social groups in rural communities become involved in decisions about environmental management. She has recently conducted research in model forests, community forests, and biosphere reserves in Canada, Sweden, and Japan.

David Robinson is an economist at Laurentian University in the School of Northern and Community Studies. He is the author of *The Economic Theory of Community Forestry*, past director of the Institute for Northern Ontario

Research and Development, and an expert on northern Ontario economic development. He teaches both regional and natural resource economics and writes monthly on the northern Ontario economy for *Northern Ontario Business Magazine*.

M.A. (Peggy) Smith is Lakehead University's interim vice-provost (Aboriginal Initiatives), formerly an associate professor in Lakehead's Faculty of Natural Resources Management. She is a registered professional forester. Her research interests focus on the social impacts of natural resources management, including Indigenous peoples' involvement and community forestry. She considers herself privileged to be a part of the growing number of people of Indigenous ancestry (Cree) who are working in the field of natural resources management, conservation, and development in Canada.

Sara Teitelbaum is an assistant professor in the sociology department at Université de Montréal. She has been conducting research on community forestry for more than fifteen years. This includes conducting the first broad survey of initiatives across Canada and in-depth case studies in three provincial jurisdictions. She has produced many educational items on community forestry, as well as an edited volume and several academic articles in international peer-reviewed journals.